Proving Ground

PROVING GROUND

Expertise and Appalachian Landscapes

Edward Slavishak

Johns Hopkins University Press
Baltimore

Printed in the United States of America on acid-free paper
2 4 6 8 9 7 5 3 1

Johns Hopkins University Press
2715 North Charles Street
Baltimore, Maryland 21218-4363
www.press.jhu.edu

Cataloging-in-Publication Data is available from the Library of Congress.

A catalog record for this book is available from the British Library.

ISBN-13: 978-1-4214-2539-9 (hardcover : alk. paper)
ISBN-10: 1-4214-2539-4 (hardcover : alk. paper)
ISBN-13: 978-1-4214-2540-5 (electronic)
ISBN-10: 1-4214-2540-8 (electronic)

Johns Hopkins University Press uses environmentally friendly book materials, including recycled text paper that is composed of at least 30 percent post-consumer waste, whenever possible.

For Stacy

Let's pause on these heights. Now you understand what I meant when I spoke of aiming higher. I was talking, it so happens, of those supreme summits, the only places I can really live. Yes, I have never felt comfortable except in lofty places. Even in the details of daily life, I needed to feel above.

—Albert Camus, *The Fall*

CONTENTS

ILLUSTRATIONS

ACKNOWLEDGMENTS

This project began in my attempt to make sense of a nearby phenomenon: the recent emergence of Centralia, Pennsylvania, as a fringe tourist attraction. Centralia appears nowhere in this book, but the recurring visits that increasing numbers of people made to this "town on fire" made me think about what remote places offer to people who work hard to reach them. What kinds of experiences do visitors seek? What kinds of proof do they take home with them? What do they actually *do* while in these spaces? Those questions kick-started an exploration of mountain visitors' practices and routines, with an eye toward how they explained themselves to themselves.

Here's how I explain myself: I wrote the book on planes, in dozens of hotel lobbies, in even more hotel rooms, in library reading rooms and museum foyers, at hospital coffee shops, in the corners of student unions, at my kids' soccer games, at restaurants between conference panels and on the way to research sessions, in three beach houses, on the shore of a state park lake, in my family's dining room, in the bedroom in which I grew up, and, above all, in an office on the third floor of a century-old academic building in Selinsgrove, Pennsylvania. I typed it on six computers and scratched it on countless legal pads; in the "notes" section of conference programs; on the backsides of fliers, book exhibit sales sheets, and elementary school cafeteria calendars; on paper plates; on the emptiest pages of the *AHR* and the *JAH*; on tiny airline napkins; in the spaces between the entries of to-do lists drafted on vacations; and in the great blue space of the Atlantic Ocean on road atlases. I stored it in computer folders titled "chaps," "NEWchaps," "REWORKchaps," "FINALchaps," and "SENTchaps"—all housed in a folder originally titled "bookbook" and then "BOOK2." At first it was all about photography. At one point, it was all about dereliction. And for at least a year, Sean Connery was central to the final chapter.

I thank the library and archive staff members who made this work possible. I especially owe great debts to Debra Basham, Elizabeth Dunn, George Frizzell, Karen Glynn, Annette Hartigan, Sarah Hartwell, Seth Jordan, Sallie Klipp, Steve Naugle, Megan Ó Connell, David Pavelich, and Michael Sherbon. I received much-needed research assistance from Susquehanna University students Sarah Kern, Caitlin Mahoney, and Chelsea Smith. Bill Hart and Susan Shoemaker helped at the right time with George Masa material, and Linda Ries made me think about Horace McFarland in a new way. I thank my colleagues in the History Department of Susquehanna University for shaking up the project by asking the right questions at the right time. Any claims to expertise that I might stake have been fostered by the comments of readers for the *Journal of Social History* and the *Pennsylvania Magazine of History and Biography* and by colleagues in the Agricultural History Society; the American Society for Environmental History; the Appalachian Studies Association; the Canadian Mountain Studies Initiative; the Identity, Conflict and Representation Research Centre at the University of Derby; the Pennsylvania Historical Association; the Society of Appalachian Historians; and the Southern Forum on Agricultural, Rural, and Environmental History. Working with Johns Hopkins University Press has been a delight. I thank especially my editor, Elizabeth Demers; my copy editor, Jeremy Horsefield; and my anonymous readers.

Finally, I thank my family. Unlike them, I'm not Appalachian born, but they don't hold it against me.

Proving Ground

Introduction

Marion Pearsall had been working for three weeks in Leslie County, Kentucky, before she had her epiphany. The University of Kentucky anthropologist arrived in the county seat of Hyden in June 1959 to conduct a "social anthropological study of health and culture" in the Appalachian Mountains. She made a few contacts, rented an apartment above a store, and settled into observing locals as they went about their lives. Then, on July 2, it hit her: the people who lived in and around Hyden had a topsy-turvy system of direction that went beyond mere familiarity with landmarks. They *experienced* their surroundings differently than she did, in a way that made the act of giving directions to an outsider a confounding task. She typed her revelation the next morning in her field notes: "finally got the 'up' and 'down' and 'lower' and 'upper' terms straight today. The county from Hyden south to Harlan County is the upper part of the county; from Hyden north to Perry County is the lower part of the county." She continued in her notes, explaining the twist to herself as a remnant of county residents' immobility and isolation until the recent past. People who lived in the northern (lower) end of the county had always thought of themselves as connected by rail lines to the relative hustle and bustle of Perry County, while the residents of the southern (upper) end of the county did not think of themselves as connected to much beyond their own communities. Rufus Fugate, an agent for the University of Kentucky Extension Service, told her that "up," "down," "over," and "across" were the essential, if counterintuitive, concepts there. This peculiar method of bearings seeped into other corners of life. Pearsall noticed that a piano teacher on the western edge of the county instructed students to play "down" the scale when she wanted them to do the opposite.[1]

In some respects, Leslie County, Kentucky, seemed like a place where the world was turned on its head. Up was down and down was up. But in other ways, the communities of Leslie County were just like communities throughout the United States. For Pearsall, a medical anthropologist interested in the ways in which social groups developed strategies for health and welfare, the area seemed like an ideal case study for human ingenuity and frustration. The mixture of familiar and unfamiliar marked the eastern haunts of Kentucky as a place to make intellectual and professional headway—if only the visitor could make the experience of such a place relatable to others. Pearsall believed that she had figured out something fundamental about the people of Leslie County on July 2. In the years that followed, she worked to convey that insight to colleagues. She wanted them to know what she knew and to see how she knew it.

A half century later, we live in an age in which old models of expertise have been trumped by do-it-yourself mind-sets, an uneven democratization of technology, and expanded access to information. At the same time, other traditional types of expertise have survived, especially those that depend on physical mobility and access to build an aura of authority. Reaching remote sites and gaining entry to places that are literally or figuratively closed still matters in fields as diverse as terrorism analysis, astrobiology, and urbex (urban exploration). In this volume, I examine moments like Pearsall's epiphany in Hyden as examples of a pervasive trend in American professional communities in the twentieth century. A wide variety of occupational and recreational groups used the landscape of the Appalachian Mountains as the field in which they staked their claims to experience, innovation, and, above all, expertise. Time spent in the mountains, in the guise of work (or play that seemed a lot like work), could distinguish oneself from one's peers by demonstrating an ability to solve problems or master the unknown. This phenomenon paralleled the general quest for status elevation which marked much leisure travel in the United States since the early nineteenth century, but it differed in the ways in which the travelers appealed for status. They were less interested in displaying their ability to enjoy conspicuous leisure travel than they were in appearing to have untied knots that were central to their disciplines. The Appalachian Mountains became a proving ground for preservationists, planners, hikers, anthropologists, and photographers, among others, all of whom conducted core-to-margin (usually city-to-woods) trips that ended up back in the core again. The relationships between landscapes and visitors relied on a pair of common aspects that gave these journeys a status-building quality for outsiders as they shared their experiences.

First, the mountain visitors played on the well-known narrative of isolation which had emerged to explain Appalachia by the early twentieth century. Scholars have shown since the 1970s that the isolation of the eastern mountains was overstated. Mountain communities and even the most secluded farms were connected to regional and national networks of exchange, politics, and culture in the nineteenth century. By the 1920s, decades of elite travel, the dissemination of mail-order catalogs, a wave of road construction, the growth of a regional lumber industry, and early missionary efforts had linked the mountains to lowland cities and infrastructure. Yet the power of the isolation theme was profound for many commentators who viewed Appalachia from afar, and it could bestow legitimacy to a trip through the highlands by emphasizing the difficulty or mystery surrounding such travel. One cultural geographer has described this view of Appalachia as a longing for a "land of misty ridges, cozy coves, and hidden hollows." Misty, cozy, hidden—these labels connote the unknown qualities that travelers ascribed to the lands and peoples of the mountains. The main figures I explore used the isolation theme to promote their own skill at breaking down boundaries or acting as liaisons. Their trips, they argued, were in the name of a greater good.[2]

Second, visitors like Pearsall presented the worlds they found in the mountains as innately oppositional. The wild environments and wary people tucked away in northern Pennsylvania, western North Carolina, or eastern Kentucky stood as counterpoints to the presumably monotonous experiences of their colleagues and constituents. The task of the visitor who wanted to accomplish something profound was to meet the contest head-on and deal with the contradictions of life on high. This impulse was sometimes presented as a back-to-the-land-style recuperation from the pressures and pace of modern life through authentic exertion. In other venues, it became an aesthetic or philosophical break from accustomed ways of viewing and thinking. The individuals studied here shared the belief that the Appalachian Mountains offset the familiar, anodyne life of industrialized, commercialized, and urbanized America. The travelers might have praised or condemned what they found on their trips, but they made it clear that facing the struggle mattered; it gave them a helpful set of credentials.[3]

From Tenderfeet to Experts

The process of proving oneself on the ground while far from home turf is part of what one historian has called the "long and dense history of working out our identity through the environment." The identities at stake in this process have been both individual and corporate. Somewhere between the lone hiker finding herself on the trail and the nation as a whole discovering its value in the majesty of natural wonders lay groups of people, bounded by occupations or affinities, who relied on outdoor experiences to gain the working knowledge necessary for their careers or passions. Scholars have studied the physical performance of travel as a social act that ends with travelers narrating their trips for friends and family. Slide shows, picture albums, and souvenirs have united those who traveled with those who remained behind by impressing upon the latter the movements of the former. People experience leisure travel as a process that includes pre-trip anticipation, the travel itself, and post-trip deliberations. When considering travelers whose destination was the very basis of their professional lives, however, the performances focused on an alternate set of links to the people "back home." The mere achievement of going elsewhere was not enough.[4]

What mattered more to the people I study here were the mechanical details of what they did in the mountains—the practices and performances they maintained while carrying out their planned excursions. Their engagement with landscape through technology and the methods of their disciplines became the currency exchanged for expert status. Throughout the book, I treat landscape in the style of the philosopher Arnold Berleant, who understands it as a perceptual unit that is sensual as well as visual, enveloping as well as distant. Landscape here is not limited to a pretty scene, but is instead a setting for human activity that "engages the entire sensorium." This is in line with cultural geographers' assertion that humans experience the meaning of places through (and not in front of) space; there is an embedded quality to engaging with nature that often gets lost in translation. This understanding of landscape centers my analysis, because the figures featured in this book hoped to translate the sensations of Appalachian places faithfully. Those sensations could be just about anything: the coarsening of one's body after dozens of climbs up rugged mountains, the colors on the horizon when the sun set behind distant pine trees, or the rattles and bangs echoing around a backyard as mountain residents fixed a decaying car. In each of my case studies, the translation offered by visitors had to account for time, space, memory, and feeling. Just as a horizon is not mappable but is open to perception, these agents believed that their personal experiences in the woods could be rendered useful to others. Each of them perceived their time in the mountains as time *embedded* in the mountains, and they portrayed that condition as the feature that distinguished them from pleasure-seekers or idle ramblers. They had work to get done, and it was through the thickness and messiness of Appalachian landscapes that they built their résumés.[5]

My mountain travelers were heirs to the middle-class occupational strivers that historian Robert Wiebe scrutinized fifty years ago. Members of the turn-of-the-century professional classes carried confidence and prestige with them as they pursued assorted activities toward disparate goals. What joined them was a collective infatuation with their own talents; members of Progressive Era professional groups asserted that they had developed unique skills to solve the nation's problems. Wiebe's notions of the "shared mysteries of a specialty" and universities' role as "outposts of professional self-consciousness" help illuminate the expertise impulse. Those factors allowed elites to see themselves not as a social class but as groups formed by preparation and experience. By achieving a working knowledge of a field, one could gain access to a network of similarly equipped workers, but applying and even extending that knowledge in new ways was expected. If the first generation of emerging professionals built a "new structure of loyalties" around occupational concerns, then subsequent generations tried to revamp the structure frequently. Appalachian marginal spaces were useful in that process; these were spots in which one could make discoveries.[6]

Landscapes are technical, political, and emotional. They are governed by power relations, property laws, and natural forces, yet they are never finished or stable. That indeterminate quality allowed practitioners of many kinds to wedge themselves into the scene. Each of my figures was intent on getting peers to recognize his or her expertise through landscape. In other words, they hoped to show that they were not "tenderfeet." The tenderfoot loomed large in the outdoors movement of the early part of the century and was discussed most by those who offered advice to prospective woods travelers. Tenderfeet came in different styles, but the leading signs of inexperience were entering the wild outdoors without the mental, physical, and material equipment to thrive or relying too much on an excess of goods. Tenderfeet did not know what to do in wild situations; they could not find comfort in rough surroundings; they did not handle surprises well. Although tenderfoot status is a simplified framework for studying experience in the mountains, it is helpful in its emphasis on the expectations heaped on those who traveled. For the visitors, there were obligations that they had to meet to succeed in Appalachia. They were connected to networks of like-minded people *beyond* the mountains much more than they were connected to people *in* the mountains. Indeed, the first three examples studied here suggest that the mountain travelers went out of their way to avoid or ignore the people living in the highlands. The scenic preservationist J. Horace McFarland, the regional planner Benton MacKaye, and the founders of southern hiking clubs preferred an outdoors that was uncluttered by people; they spent their time in mountain spots that were high enough and remote enough to make it more likely that they would run into bears than humans. Those local people whom they did encounter they could quietly omit from their accounts. Their sense of duty in the woods was oriented toward mentors, colleagues, and clients beyond Appalachia.[7]

In the second half of the book, when the mountain visitors actually engaged with local residents in a purposeful way, the ties of obligation became more complicated. Marion Pearsall and the photographer William Gedney went to Appalachia to be with people, to study them intellectually, behaviorally, and aesthetically. Both visitors came to believe that the relationship between residents and the land was the key to understanding their lives. Unlike the first set of travelers, for whom shedding tenderfoot status meant proving their achievements for people who were unacquainted with the region, Pearsall and Gedney first had to establish rapport with locals. They had to observe them and allow themselves to be observed, engaging in a mutual gaze, with all of the assumptions and prejudices that influenced such interactions. Pearsall and Gedney recognized this early in their stays and used this insight to claim a special ability to decipher eastern Kentucky.[8]

A consistent element of the tenderfoot concept was not only being new to the wild but also being aware of how new one was. In popular fiction at the turn of the twentieth century, tenderfeet usually suffered a comeuppance, when rugged woodsmen or leathery ranchers showed them just how unprepared they were for the "real" outdoors. Tenderfeet were ever conscious of their own vulnerability—their exposed skin was not yet protected by the callus of expertise. A welcome by-product of becoming a seasoned agent in the wild, then, was an end to these anxieties. Confidence and experience went hand in hand, and when wielded by Appalachian visitors, this confidence tended to become a license to intervene. The organizations and networks that my subjects represented had vested interests in altering, "saving," and interpreting the mountain region. They instilled in their practitioners certain ways of seeing landscape. McFarland and his fellow City Beautiful advocates saw the outdoors as an aesthetic tableau, brimming with "color, texture, mass, line, position, symmetry, balance, tension." MacKaye saw the outdoor world as a system, an interconnected series of processes that could be read through the surficial clues of trees, soil, and rocks. The members of the Smoky Mountains Hiking Club and Carolina Mountain Club perceived the Appalachian wild as a challenging habitat, in which they, themselves, were the creatures striving to adapt and survive. Pearsall saw her focal points in eastern Kentucky as rich places, with the physical environment influencing the local details that made life in particular communities meaningful. Finally, Gedney viewed landscape as a history of accumulations and a testament to the fact that "life must be lived amidst that which was made before."[9]

Serving as surrogates for their disciplines, these nascent experts brought their baggage with them. They viewed Appalachia from the angle of the city, sometimes explicitly, as in the cases of McFarland and Gedney, and sometimes subtly, as with MacKaye and the Smokies hikers. They overemphasized the wild, untouched aspects of the landscapes they found, perhaps to make it an appropriate "field for the exercise of power." They fit very well into literary ruts worn since the 1880s. Outsiders thinking about Appalachia have presented a pervasive sense of difference rooted in temporal and topographical cues. The invention of Appalachia as a constellation of cultural references has relied on the visual relationship between pathetic human figures and an unnerving physical environment. For example, the "hillbilly in a holler" image typified one style of documentary photography in the 1960s, meant not to mock the subjects for their destitution but to criticize mining companies for preying on mountaineers' land and federal and state governments for turning a blind eye. Anthropologist Allen Batteau argues that images of Appalachian poverty moved beyond pathos in the sixties by creating a context for destitution in the ruin of the landscape and the years of neglect. The terrain of this particular Appalachia was both touched and, weirdly, untouched—thoroughly worked over through decades of timber and coal extraction and yet undeveloped in terms of infrastructure and commerce. In a nation that prided itself on technological development and a forward-looking stance, Appalachian communities were easy targets for the backward label. Twentieth-century visitors interpreted poverty in the mountains as a stubborn economic inertia on the part of its residents. The experts studied here were intent on providing a regional outlook through their efforts. As much as they focused on—and even created—specific places through their fieldwork and written summaries, they attached those places to the larger structure of Appalachia.[10]

The concept of the region is more rhetorical than descriptive; when wielded by outsiders, it is meant to simplify the environmental and cultural complexity of numerous places to make them combinable. My subjects were all convinced that there was an identifiable Appalachia to be experienced. We can see in their efforts to visit and narrate the mountains the process that helped sustain the notion. Just as special places like state parks required massive amounts of effort to produce a vision of seemingly pure nature, the region of Appalachia required a great deal of practical and rhetorical labor to remain a viable construct in American life. When MacKaye stood in the mountains of Tennessee and felt himself connected directly to Massachusetts, or when Pearsall explained her activities in Leslie County as a case study of the southern mountains, they implied that geography, topography, and altitude mingled to influence a common culture-in-place. I proceed from the assumption that studying the region concept reveals the conveniences it provided to certain groups that oscillated between the interpretation of specific places and the extension of that interpretation to the wider area.[11]

Interventions

As geographer Robert D. Mitchell notes, it is through the analysis of episodes such as these journeys that the physical area known as the Appalachian Mountains and the "distinctive social conditions in place," known as Appalachia, merge. For forty years, Appalachian Studies scholars have been thinking about the interactions between mountain residents and nonresident "intervenors." David Whisnant established an approach to these figures which has continued to influence treatments of people from outside the mountains who moved in and (usually) moved back out. Whisnant's most lasting contribution to the concept is his analysis of intervenors' power, which has allowed them to "define what the culture *is*, to normalize and legitimize that definition in the larger society, and even to feed it back into the culture itself, where it may be internalized as 'real' or 'traditional' or 'authentic.'" Intervenors' actions relied heavily on the claim that the residents of Appalachia had allowed themselves to be bypassed by change and improvement, necessitating the aid of outsiders to drag the region into the twentieth century. Collected in the phrase "the politics of culture," the power relationships central to outsider/insider structures have given several generations of scholars plenty to study in the eastern highlands. The general model that they have constructed features progress-minded outsiders, schooled in the helplessness of Appalachian residents, embarking on political, cultural, infrastructural, economic, and agricultural programs to solve problems. Their efforts might have improved some cosmetic (or even significant physical) factors of life in the mountains, but their involvement ultimately fostered the appearance of progress while masking the deep economic and political exploitation that maintained the status quo.[12]

The ubiquity of the intervenor narrative in Appalachian Studies has meant that scholars have often disregarded the actual practices of visitors while foregrounding the unintended consequences of their actions or the "energy sink" that they created by launching seemingly apolitical cultural programs. The scholarly take on intervenors has run the gamut from describing them as "well-intentioned others," or depicting them as forces of "using and consuming" rather than "connection and mutuality," to presenting them as the "seers and fondlers" who slowly corrupted what they traveled to experience. Intervenors have also come to stand in for the mass media's negative portrayal of Appalachia, and indeed, several of the figures I follow here attracted audiences that were large enough to be labeled "mass." Intervention has become shorthand for actions made by people who were convinced that they "got it," but did not—or could not. When the West Virginia activist Judy Bonds spoke of twenty-first-century American views on how people should act, she stressed the judgment that a monolithic society heaped on the mountains: "We don't put on a show, don't put on airs. We're unhomogenized. We're not Photoshopped. We're unretouched. We're *real* people. And I don't think mainstream America likes real and stark and raw." In the analyses of Appalachian Studies scholars, intervenors can come across as agents of inauthenticity and senseless power.[13]

The intervention discussion in Appalachian Studies mirrors the wilderness debates in environmental history. These debates consider whether the mostly urban preservationists and conservationists of the first half of the twentieth century—and wilderness advocates since that time—have done more harm than good by fostering the modern concept of wilderness as a pristine zone apart from human affairs. Some critics of the wilderness concept have argued that it was the product of attempts by European colonists to push Native Americans from the land. Others have shown how the class biases inherent in wilderness preservation and outdoor recreation in an industrial age made the establishment of public parks and forests a well-funded move to deliver the outdoors to urban elites. William Cronon pointed out the harm of this lingering valorization of distant, human-free wilderness at the expense of the everyday coupling of people and their environments. Expending time and effort to save the pure mountain peak with its virgin forest runs the risk of neglecting the urbanized and commercialized, yet still natural, worlds in which many environmentalists live. There is wildness all around us, Cronon cautioned, "if we only have eyes to see it." Whatever the specific criticism, historians of the twentieth-century United States have shown that wilderness was only one of several ways of modeling the "proper relation of humans and nature," and that it was the power of the state and the wealthy—especially their ability to impose their will on locals—that made wilderness triumph as a useful concept. The point of much recent environmental history has been to unravel that political process. Geographer Edward Casey reminds us that there are "no First Places that altogether withstand cultural perversion and specification." Specification is one thing, but perversion is another entirely.[14]

Proving Ground studies the specification, whereas other scholars have tended to interpret the cultural and environmental perversion. A different kind of book could certainly be written about the people and places I consider. It would be a book in which Eagles Mere, Pennsylvania, became a superficial preening ground for the fashionable set of the Mid-Atlantic. The Georgia-to-Maine hiking trail across Appalachian peaks and valleys would be the indulgence of weekend warriors from nearby cities. The Smoky Mountains would be the place where state and federal agencies pushed mountain residents from their land, several generations after logging companies cut over all but the trickiest slopes. Leslie County would become the place where that professor from Lexington asked her questions and tried to act like a local. Big Rock might be the spot where the New York art photographer replaced one set of Appalachian stereotypes with another—and benefitted when his colleagues chose not to notice.

Yet if we think about Appalachia as John Alexander Williams does—as a "screen upon which the writers, artists, and savants of five or six generations have projected their fears, hopes, regrets, and enthusiasms about America present and past"—then there is much to learn about how "fears, hopes, regrets, and enthusiasms" can be experienced through landscape. A way out of the woods is to consider (without celebrating or excusing) the motivations of the intervenors more closely. The trips into the Appalachian Mountains were inspired by what anthropologist Tania Murray Li calls the "will to improve." Li argues that centralized governments and nongovernmental groups alike understand peripheral land and populations as opportunities to perform acts of improvement. The figures discussed in this volume did not believe in saving Appalachia through the "education fix," the "federal fix," or the "grassroots fix." They worked to improve their own fields and their standing within those fields. Individual careers, professional fields, and political regimes are formed around the problems that intervenors articulate and then work to solve. Central to the process of articulation and solution is what Li calls "rendering technical," a series of steps through which practitioners gather data, theorize causes and effects, outline possibilities, shape the vocabulary, and, all the while, assert their expertise. Rendering technical is what happens when people with experience are given the time and resources to ply their craft; it is also when the symbolic violence gets done. Their maps, charts, field notes, policy proposals, photographs, and trip reports establish their relationships with environments as authoritative. These devices tell outsiders that the experts know the lay of the land and the cultures, perhaps even implying that they know these things better than residents.[15]

Insiders and outsiders are tricky types. At the most literal level, determining insider or outsider status is a matter of plotting places of birth or residence on a map. At a more perceptual level, the process could involve taking subjects' statements and behaviors into account and/or charting their familial and affinitive relationships for evidence of connection to a particular place. A problem quickly emerges in this process; if one is determining who counts as an insider, exactly what is it that the individual is inside? The next three chapters examine people who assumed that it was difficult, even unnatural, to feel at home in the Appalachian woods. Their careers hinged on their ability to offer instruction to readers and listeners who wanted to know how to "do it right." McFarland, MacKaye, and the Smokies hikers were careful to depopulate the mountains in their presentations. Their instruction positioned audience members as outsiders, but they, the experts, were not necessarily insiders. Hiking groups argued that the exhilaration they felt at mountain overlooks meant nothing without the advanced proficiency they gained in the mechanics of hiking. The clubs sought a specific form of insider status: an embodied engagement with extreme material environments. This was a practical competence that might serve them well in the Smoky Mountains but was not limited to any one place. Rather than being an insider, it was being professionally skilled that meant so much to these actors.[16]

The last two chapters show what happened when mountain visitors acknowledged the presence of people on the ridges and in the hollows and assumed that with patience and sensitivity a visitor could become at home within the Appalachian Mountains. But even then, the home to which they belonged was not a mountain community or a regional spirit. Brooklynite William Gedney lived with the Cornett family in Big Rock, Kentucky, long enough that the Cornett matriarch asked him, "Bill, why are we so attached to you? You feel so close somehow, like you are close kin to us." Yet Gedney wrote about the Cornetts and their neighbors with a detachment that did not suggest any intimacy. As much as they were intent on mastering the conditions in the mountains, each of these figures was open to the possibility of being overcome by what they found. They were flexible enough to acknowledge that they could never learn everything there was to know about their destinations, and thus the expertise that they brandished to their colleagues was less a knowledge *of* Appalachia and more a knowledge *through* Appalachia. Geographer David Crouch has called this lay knowledge—accumulated slowly through experience and repeatedly rewritten from fragments of memory. They sought an expertise that happened in the mountains because of what the mountains meant to so many who did not live there. In these mobile practices, there were always multiple identities of which an individual could be on the inside or outside. It was not always a matter of belonging to a place.[17]

Read the forewords or introductions to most books written about Appalachia by someone who was raised in Appalachia, and you will find writers going out of their way to vouch for the authors' roots. Being a son or daughter of the mountains seems to matter as a credential for accuracy of insight. The people studied here were geographic outsiders who told stories to readers and viewers who also did not live in the mountains. But this is not a book about outsiders; it is an example of what Richard White has detected as a scholarly shift into "discourse, story, and narrative" about "hybrid" landscapes. Appalachian Studies scholars have recognized that an interpretive overreach can turn all nonresidents into "inauthentic" figures at best and diabolical schemers at worst. The hybridity emerges here in the collapsing of various types of media into expert practices (Li's devices for rendering technical) and the collapsing of insiders and outsiders into ambitious individuals attempting to juggle professional, artificial, embodied, and altruistic identities in particular places. Scholars have recently challenged the assumption that belonging is something that accumulates and then persists. Approaching senses of place as "fluid, changeable, dynamic contexts of social interaction and memory," these writers suggest that attachments to specific landscapes come and go, changing over time within individuals and through the images that circulate within wider cultures.[18]

The hybridity of belonging parallels the complex histories of wild spots. The contradictions and ironies of the environmental movement in the United

States in the late nineteenth and early twentieth centuries are well rehearsed. Historians have stressed that tourism promotion made people appreciate undeveloped areas with new gusto, yet the popularity of the wild then increased fears that such areas were under threat. Consumption of scenery and outdoor recreation worked seamlessly with the conservation efforts that seemed to discourage them. Tourism driven by aesthetic responses to scenery often relied on a convenient forgetting—a denial of the fact that government management, industrialization, and mass production technologies produced "untouched" wilderness. Since Cronon's call for a more explicit recognition of the politics of wilderness, historians have attempted to show evidence of both the wild in the city and the city in the wild. Together, they have created a continuum approach to outdoor promotion and reform movements, in which nature and culture are always present to varying degrees.[19]

Hybrid landscapes are also occupational terrains; lay knowledge becomes useful to individuals as it accrues within specific fields of work. My subjects' thinking about the landscape of Appalachia was influenced by their colleagues' interests and the fixations of their disciplines. This is not a new discovery; even if scholars have neglected the occupational stakes of movement to and through the Appalachian Mountains, they have certainly demonstrated that intellectual approaches to undeveloped areas are discursive formations. And many discourses are directed via professional cohorts. Tourism promotion in the United States followed more than a century of incremental work by astronomers, botanists, and philosophers, among others, to frame the outdoors in new ways. Historian Roderick Nash showed how the Romantic ideal of wilderness as a positive force was fostered by "writers, artists, scientists," and others who shared lives "surrounded by cities and books." Reacting warmly to the wild became a way to mark oneself as studied and sophisticated, schooled in new aesthetic principles and naturalist explanations. By concentrating on the privileges of the narrating elite—what Nash called "the vantage point of comfortable farms, libraries, and city streets"—it is easy to disregard the occupational dimension of those principles and explanations.[20]

My focus on Appalachian visitors' practices relies on close attention to the messy material of human encounters with landscapes. Crouch stresses that spaces of travel combine the material and the metaphorical. Bodies sense spaces materially, he writes, but "space and its contents are also apprehended imaginatively, in series and combinations of signs." Taking this insight seriously means refusing to relegate Appalachian spaces to inert surfaces and packaged spectacles observed by those who were passing through. Although I examine representations and consider their ability to influence panoramic perceptions, I place equal emphasis on the "close-up" spaces in which people snapped pictures, hiked, drove, climbed, interviewed, slept, or ate.[21]

Organization of the Book

Anthropologist Kathleen Stewart imagined writing a fifty-thousand-word description to relate what it felt like to do fieldwork in Amigo, West Virginia, in the 1980s. Devoting such space to the sensory experience of one Appalachian place, she dreamed, might get the reader beyond "you are there" realism to arrive at an intense feeling of the local. The protagonists of this volume shared this fantasy of indexing Appalachian environments, even as they subsequently turned them into abstract landscapes that defied detail. On the one hand, they piled detail upon detail to describe their movements up mountains, through hollers, and down creek beds. They labored to impart a thick, multisensory knowledge of their spots on the map. On the other hand, they could not help themselves—in fact, did not want to help themselves—from translating Appalachia into a panoramic spectacle. They experienced the mountains with a blend of the "generalizing cultural visions and environmental interpretations" typically attributed to outsiders and the "localized knowledge and experience" said to be the domain of insiders. The book follows their attempts to scrutinize and solve in roughly chronological order.[22]

Chapter 1 studies the mountain work of J. Horace McFarland, the leader of the American Civic Association between 1904 and 1924. Based in Harrisburg, McFarland considered northern Pennsylvania's Appalachian wilds as his key occupational realm. McFarland created a strategic position for himself at the crossroads of environmental politics and genteel aesthetics. Using the professional languages of botanists, photographers, and Progressive preservationists, McFarland applied City Beautiful reform techniques to shape the experiences of mountain tourists. To promote the twin concepts of "sightliness" and "woodsiness," he recast his love for cars and cameras into a system of picturing the outdoors. His takes on the wild were as much about those technologies as they were about the trees—and he certainly loved trees. McFarland's expertise was grounded in his fieldwork: annual excursions on wheels and on foot throughout the hinterland to the north of the capital.

Chapter 2 turns to another woods walker, Benton MacKaye, as he designed an ambitious experiment in regional planning along the spine of the Appalachians. With occupational bases in Massachusetts, New York, and, later, Tennessee, MacKaye used his training in forestry and his commitment to socialist principles to devise the Appalachian Trail as a new mode of experiencing the natural world. Although underemployed throughout the 1920s and 1930s, MacKaye enjoyed strong connections to an influential group of urban reformers and planning officials. They encouraged his work, and, in turn, MacKaye couched his thoughts about the Appalachian wilds in the language of their professional networks. Before and after the publication of his trailblazing *The New Exploration* in 1928, MacKaye relied on time spent in the eastern mountains to "ground" his abstract concepts of social and natural forces colliding over time.

Chapter 3 moves south to the rugged border of Tennessee and North Carolina, where the Smoky Mountains Hiking Club and the Carolina Mountain Club emerged in the 1920s to organize the experience of mountain enthusiasts. The founding members were urbanites who traveled from regional cities like Knoxville and Asheville or modest towns like Maryville, Tennessee, to climb the steep slopes of the Appalachians. For them, strenuous hiking was a way to extend the qualities of their working lives into the wilderness. Their professional lives, spent in banks, law offices, classrooms, and chambers of commerce, informed their serious style of enjoying the outdoors through teamwork, diligence, incremental achievements, and hierarchical communication. In the process of mapping and indexing the Smoky Mountains in the 1930s, club members created a quasi-professional recreation in an area that had not seen such organized outdoor leisure.

Chapter 4 follows the "rediscovery" of Appalachia by the national media after 1950 and filters that phenomenon through Pearsall's work in medical anthropology. As out-migration, labor strife, and federal funding produced a political culture that cast mountain residents as feeble victims who could not help themselves, Pearsall embarked on studies of remote communities in Tennessee and Kentucky. She was dedicated to action research, a mode of scholarly observation and activism that encouraged experimental forays to challenge the status quo in a given arena. The national media and federal policy makers depicted the "problem" of Appalachia in technical and material terms—as a matter of poor soil, poor traffic systems, poor water management practices, and poor lifestyles. Pearsall considered how small tweaks on the ground (rather than massive infrastructural efforts) could improve lives in short fashion. Action research offered a menu of strategies for locals to adapt to their own conditions, and Pearsall's insights about her field sites thus reveal her budding sense of the possibilities of social science.

Chapter 5 analyzes the Appalachian imagery of William Gedney between 1964 and 1972. In Gedney's photographs of eastern Kentucky, people, landscape, and economic systems overlapped to produce a flurry of surfaces and signs. The city boy traveled to the heart of the central Appalachians at a time when journalists and documentarians had made this coal and lumber country a regular feature of current affairs. Gedney stated bluntly that he was not a "social problem photographer," yet his Kentucky photographs presented closed-in places where unemployment, dilapidation, and dereliction reigned. Gedney's street photography style adapted well to places where streets were dirt tracks through the trees and buildings were makeshift cabins weathered to a dull gray. His Kentucky work was the centerpiece of his own career, propelling him forward among New York photographers and arts patrons. In the manner of Walker Evans, Gedney brought an urban eye to rural settings and hoped that individual subjects might speak about the state of the nation.

The characteristic act of observing while traveling, writes John Taylor, is insisting on difference. This book sets out to show that these five episodes of expertise in the making fit into a pattern of thinking about the Appalachian Mountains in which grappling with difference demonstrated the worth of the traveler. These were searches for meaning, and yet they promoted a looseness of interpretation that shied away from definitive answers, even if they were supposed to be authoritative. What mattered more was convincing others that they had been true to a moment on the margins. For each of the case studies presented here, authenticity and truth in the mountains were not stable concepts, and the players involved recognized it. Each of them wanted an Appalachian experience that was mitigated or translated by technology. Each of them considered authenticity in a way that could be described as consumerist; it was something that one pursued as a part of a regular urban lifestyle. The authentic was not fundamentally different from the artificial—they were part of the same story. The stories that follow consider the nature of the proof that these people wielded to cast themselves as experts on some of the most marginal spots in the country.[23]

CHAPTER ONE

Expert Vision

J. Horace McFarland in the Woods

When he lectured before public audiences in the 1910s, American Civic Association (ACA) president J. Horace McFarland liked to talk about a recurring scene from his professional life. In the scene, McFarland sat at his office desk in Harrisburg, drafting memos to fellow City Beautiful reformers, reading legislative reports concerning scenic preservation, and meeting with his many visitors. All the while, his eyes flitted to the photographs on his office walls, images of wild scenery which he had taken on trips into the mountains of northern Pennsylvania. The noise and motion of the city dropped away when he looked at the trees, streams, and distant ridges. The sudden stillness of the office and the crispness of the photos carried him out of the capital and into the wild. No, he could not make the trip just yet. He was needed in the city, and so the woods must wait.

This compromise became a defining vision for McFarland in the first three decades of the twentieth century, despite the fact that he was hardly a prisoner in Harrisburg. He enjoyed the company of an active circle of friends and colleagues. He found places of great natural beauty in the city's parks, in the renovated riverfront, and in his own garden, filled with roses and peonies. Yet it served his interests to present himself as biding his time while gazing at framed mountain scenes. The idea of an urban professional pursuing his life's work, while constantly hemmed in by the white-collar world, was central to his script. Yes, he found ways to break out. As he reminded those assembled at a forestry conference in 1929, state highways allowed him to work in the morning, eat lunch at home, and still savor the "solemn shade" of the wild before dinner. Likewise, when he relocated for weeks of leisure in the mountain resort of Eagles Mere, in Sullivan County, McFarland knew that he could return quickly to Harrisburg if need be. Sequestering oneself in the mountains was a delight, yet someone of McFarland's activity could not abandon professional duties. He encouraged businessmen to take their families, noting that patriarchs could work in New York or Philadelphia in the week and steal away on the weekend. He needed the city, he loved the city, and yet he spoke often about fleeing the city.[1]

McFarland's rhetorical sleight of hand was part of his effort over several decades to convince others that he had the credentials to guide them in thinking about nature. Although he became quite well known during this period, he was a rather unremarkable scenery advocate. His pronouncements on preservation and aesthetics echoed the insights of contemporaries who wrote for the popular press or coordinated their energies through local clubs. His boosterism was on par with a slew of individuals who were invested in particular resort destinations and getaway spots. He was an exemplary elite man of his era, right down to his disdain for hot dog stands. In terms of his public life, McFarland's defining characteristic was his drive to be a tastemaker. His work between 1900 and 1925 reveals a basic mechanism through which figures like him crafted a reputation for themselves as in the know. McFarland's career is a case study in the invention of expertise—an expertise wielded in the name of reserving places like Eagles Mere for aesthetic uses. The back-and-forth between the mountains and the city was the heart of his outdoor knowledge. For an audience of city and town residents linked by the era's religious, civic, and commercial institutions, McFarland offered a vision of nature that was always mediated by technologies of perspective and access. To him, cars were a functional way to reach the mountains and the platforms from which city dwellers made the transition into wild settings. Cameras were both a useful means of recording scenes for later recollection and devices to train aesthetic experiences. Through these technologies he translated the mountain woods for an urban, middle-class audience.[2]

McFarland's example suggests that the creation of a wilderness aesthetic was not only a power play used to control mountain forests but also a strategic move within the handful of professions that collided to form environmental advocacy campaigns. He was a public relations strategist before such a position formally existed. In their study of PR efforts in the decades before the organized profession emerged, Margot Lamme and Karen Russell found that social reformers, religious leaders, and government officials all used the techniques that would eventually become the bedrock of corporate PR departments. Across time and place, note Lamme and Russell, the motivation behind such PR work was to make money, to recruit members, to establish legitimacy, to rally voters or consumers against something, or to advocate *for* something. Much of the history of environmental reform in the United States has understood advocacy as the prime motivation for outreach, while paying little attention to the third function: activism's effect of vouching for the expertise of those involved. McFarland was often an overt activist, but even when he was not endorsing a specific policy or appropriation, he relied on the urgency of others adopting his way of seeing. His authority was not a given, especially not within the specialist circles of botany, photography, and forestry; he had to earn it slowly, adapting his experience in the publishing and promotion trade to the business of telling others how to experience the outdoors.[3]

If McFarland was a star of the cultural elite, he was a nervous star; he did not assume that he had an automatic audience willing to listen. Instead, he relied on a type of expertise that stressed practical experience in the woods. The self-styling of expertise was certainly a manifestation of cultural elitism, but the technique with which McFarland tried to secure his expert status is worthy of close attention. If McFarland was to stake a claim for himself as an expert in Pennsylvania's wildness, he could do so by extending to the mountains two sets of skills that he had developed in the city. First, he applied techniques of spatial reform and beautification to structure experiences for city dwellers who were unaccustomed to the highlands. He took the serious business of landscape (and social) engineering and bent it to the recreation needs of urbanites. Second, he recast his lifelong zeal for automotive and photographic technology, creating a way for amateurs to experience the mediated wild. He seized these tools of leisure to develop a system of contemplating the outdoors. McFarland worked to establish his expertise in both aesthetic appreciation of nature and technical mastery of the newest means of engaging with it. All the while, he staked a claim to these forms of expertise through tales of his excursions on wheels and on foot several hours north of the capital.

The Lobbyist

McFarland was born in the central Pennsylvania town of McAlisterville in 1859 to a printer and nursery owner. His father published what McFarland later called a "belligerent temperance weekly," and the economics of the printing business caught the youth's attention early. He combined his father's pursuits by the age of twenty, becoming a printer "for nurserymen, florists, and seedsmen." His business, Mount Pleasant Press, was lucrative enough to fund his passions for travel and photography and incorporate him into the capital's cross-pollinating civic, commercial, and cultural elite. McFarland studied plants avidly, having joined Pennsylvania's horticulture society in 1881. He published his first wave of articles on horticulture at the turn of the century, becoming *Outlook* magazine's plant and tree writer. His contributions to *Outlook* led to the publication of *Getting Acquainted with the Trees* in 1904. In the same year, he began a three-year stint writing the monthly "Beautiful America" column for the *Ladies' Home Journal.* He was a popularizer in print and a schmoozer in person, meeting as many people as he could. His knowledge of the printing industry led him into both business partnerships that spanned the nation and regular speaking engagements. He referred to his various publishing interests as "constructors of catalogues and builders of business." In 1911, he taught students in Harvard Business School's "Technic of Printing" course. His glowing endorsements of printing devices and instructional guides appeared often in the trade press. By the time he was fifty, he had established himself as a reliable authority in multiple facets of life.[4]

McFarland was a major player in a fairly small circle of beauty advocates in central Pennsylvania, and it would have surprised few of his friends and acquaintances that he took a leading role in the Harrisburg beautification campaign that emerged in late 1900. When the political effort made great local strides, McFarland wrote about its victories for a national readership eager to apply his model. His stock rose. His highly visible position in the movement allowed him to climb the swelling ranks of city reformers. To accompany his high standing in the printing industry, he made his second professional name as the president of the ACA, a position that he attained in 1904 and held for the next two decades. His leadership of the ACA marked him as the most visible of what historian William Wilson labeled the "organized, dedicated, and informed laymen" who drove the City Beautiful movement. McFarland was the consummate publicist, speaking before audiences big and small and writing columns in any magazine, journal, or newsletter that would have him. By the time of his death in 1948, he was best known for his campaign to save the scenery of Niagara Falls from industrial development—and for his lifelong promotion of roses.[5]

McFarland's leadership of the ACA made him a national leader of the aesthetic wing of the conservation movement, steeped in what historian Jon Peterson considered its "genteel aestheticism and its sublime faith in the all-knowing expert." He drew criticism from self-described utilitarian conservationists for his presumably weaker stance. Yet he was never a one-trick pony. When he argued for scenic preservation in the early twentieth century, he did so in ways that were absolutely pragmatic and unquestionably strategic. McFarland exhibited what historian Kevin Armitage has called the "multitudinous cultural complexities of the conservation movement." He worked both the halls of government and the crowds at civic clubs. He used his position in the ACA to speak to municipal officials and business groups about conservation on their terms and positioned himself as their ally. When he campaigned against the aesthetic blight of hot dog stands along city streets, he encouraged thoughtful simplicity in the design of the shacks. More appealing stands would bring in more customers and harmonize with their surroundings. Instead of advocating for the abolition of billboards, "a sort of fungus on the body politic," he pushed for the establishment of dedicated commercial corridors with a lower speed limit. The landscape and the "signscape" could coexist, he reasoned, as long as advertisers realized the damage they did to sales. If they limited their exposure, they would turn more heads and make more money. Everyone would win.[6]

In other words, McFarland was an adept lobbyist, always considering what he wanted to gain in light of what people in power wanted to hear. He was not alone in these talents; the leadership ranks of the ACA were crowded with such go-getters. Geographer Terence Young observed that when the ACA focused on a given issue, its "well-positioned members, who knew how to effect change, quickly sought out and organized local support." McFarland told anyone who would listen that if government officials only viewed outdoor scenery as a "productive resource," the United States might compete with Europe for the "millions of beauty travel." He likewise targeted middle-class audiences by turning spatial reform into a prescription to save cities and stave off class conflict. His column in the *Ladies' Home Journal* was a frequent source of prodding; he encouraged readers to do everything from cleaning up their backyards and planting flowers to joining local reform clubs. He promoted the "direct economic effect of suggestion and environment" as a way to keep urban mobs in check. If white-collar urbanites refused to bankroll the repair of their cities to deflect crowding and labor conflict, they risked losing their privilege to a class war. "Parks are cheaper than policemen," he observed, and he saw adding green spaces to cities as a first step in reform. But it would be a first step only. The public was well intentioned, he reasoned, but it needed to be mobilized. His favorite tools were flattery and an overwhelming amount of supportive detail. He applauded Harrisburg's beautification campaign in 1901 and 1902 as a masterful use of local media. When the campaign started, he wrote in *The World's Work*, "each of the three daily papers in the city was supplied with carefully prepared matter to inform the voters, in a cumulative fashion." High school boys carried map- and diagram-laden pamphlets door-to-door twice a week.[7]

McFarland knew that a great many of the people answering those doors or reading his articles were women. He worked closely in the Harrisburg campaign with women such as Mira Lloyd Dock, who had studied botany at the University of Michigan and was a staple of the central Pennsylvania lecture circuit. McFarland presented himself to women as an ally—a wise counselor who knew how the world worked and believed that women had a significant role to play in it. He was a firm supporter of the municipal housekeeping brand of activism on the part of women, and it was through civic clubs, he hoped, that the housekeeping would take root. When experts like McFarland exercised social and political power in the early twentieth century, it was because they exhibited types of experience that were persuasive to the inexperienced (what historian Peter Dear calls "culturally sanctioned" experience). Experts might stand alone as *the* voices to which others should listen, but their claims to expertise are always formed within a social context and tested among peers who judge their merits and help them build their reputations. For McFarland, the "force of accuracy and the grace of clear statement" convinced others that they were in the presence of an expert. One of the peer groups that mattered most to his status were reform-minded women with organizational clout.[8]

His efforts on a local scale convinced McFarland that the public was a force that could be shaped by expert hands. "Public opinion in America is dominant," he wrote in 1908, "and when aroused, restless." The "cumulative" approach adopted in Harrisburg was something that McFarland carried with him; he assumed that his readers were following along. And yet he found a pervasive political disengagement everywhere he looked. He diagnosed it as a problem inherent in a consumer society. McFarland's goal was to mobilize professionals and use them as an "unconscious combination of militant citizens." William Wilson found that McFarland and his fellow sponsors of the Harrisburg cleanup plan were middle-aged, educated, economic elites with a firm sense of class duties. McFarland's politics were elitist; he was committed to preserving a power structure that doled out incentives to the masses. The urban professional class would be the force he used to make his ideas stick, and his rhetorical power over them would build through explorations of places they had never experienced.[9]

Getting Acquainted

It was within this problem-solving performance that McFarland worked to establish himself as an expert. Both of his audiences—the decision makers and their constituents—might embrace policy proposals that promised tangible benefits and carried the weight of experience. As McFarland noted privately in 1915, he believed that he had the "welfare of the community and the nation at heart." This was a simple way of describing the national goal that the ACA secretary Richard Watrous had announced three years earlier: an "atmosphere that makes for health, happiness, good citizenship, and material prosperity." As historian Julian Chambliss notes, McFarland forged his reputation in the Harrisburg beautification campaign. When he attempted to connect the city and the wild in the public imagination, however, his leisure time beyond the city increased in rhetorical significance. The raw material of those efforts was McFarland's time on the roads and trails of Pennsylvania's wild spots. And so his project became the translation of one man's private enjoyment of the outdoors into a general appreciation of nature.[10]

Class privilege and gender privilege were vital to the assumed translatability of his vision; they allowed him to avoid too much scrutiny and to escape the question of why he, of all people, should be heeded. His professional life was launched from business connections within the Harrisburg Board of Trade, and he recognized the concerns of industrialists and other employers. The language he used in print and in person assumed the authoritative tone of a man who dined regularly with powerful men. He was a leisurely motorist, an ambitious photographer, and a studied woods rambler, a combination that depended on a professional salary and male privilege. The mysterious aura of legitimacy suggests how the printer from McAlisterville became the president of the ACA. Dear argues that experts rely in part on an "unanalyzable residue of brute credibility" that can be maintained only through a collective willingness not to ask too many questions. Cross-examining experts to find the limits of their experience ends up weakening their overall claim to expertise. McFarland worked himself into positions in which he determined the path and focus for others. He was used to having the last word.[11]

The gravity that he needed to speak to Board of Trade members joined with the conversational, yet omniscient, manner of the advice columnist to produce McFarland's rhetorical approach. As he understood it, what was "good for America" was a mixture of commercial success and spiritual sustenance. When he titled his 1904 book *Getting Acquainted with the Trees,* it was an apt description of his goal. This was not to be dependence, nor even perhaps a close friendship, but an acquaintance. Visitors to the outdoors—read as middle-class men and women with enough aesthetic training to grasp his insights—could get just enough of an experience to want more. A reviewer in St. Louis remarked that McFarland "chatted" his way through the book, and to welcome effect. "In lieu of getting acquainted with the trees themselves," the reviewer noted, "getting acquainted with Mr. McFarland's book is fairly pleasant."[12]

Acquaintances came in many forms, but scenic roads and mountain enclaves were the two venues through which McFarland encouraged budding outdoors enthusiasts to become more familiar with nature. First, his plan was to shape the way in which motorists experienced the wild by easing them into it; rural roads maintained by the state could feature carefully managed flora to frame the road scenically. As the physical link from the city to the wild, scenic roads prepared urbanites. If designed correctly, roads could educate through the power of what he called "sightliness." McFarland's vision was a natural area that looked natural, a place that had experienced human intervention but hid it well. He liked to quote the author and photographer Wallace Nutting, who praised the Pennsylvania countryside as "never wild or terrible," but consistently pleasant. The steady motion of automobile travel allowed the state's rolling landscape to reveal itself, and McFarland believed that planned roadside nature was just as vital as the proverbial untouched forest.[13]

At his most evangelical, McFarland used the concept of the road as a "well-ordered" museum that could educate the traveler about the worlds beyond the tree line. Outdoor attractions in the eastern highlands were, for him, the "great vistas . . . waterfalls . . . [and] picturesque roads," all of which made it "worthwhile to go into the woods." As important pieces of the experience of nature, roads had to meet certain aesthetic standards. He wanted to use state funds to line highways with plant life that "belonged" in the vicinity. In a *Country Calendar* article he criticized road maintenance officials for clearing the "more delicate and beautiful" plants and leaving behind "vigorous but really unpleasant weeds." The notion of engineering naturalness along highways fit his predilection for a thin layer of human order superimposed on areas that seemed otherwise untouched. He insisted that country roads needed footpaths built next to them as a way for the enlightened, enthusiastic public to travel "between farm and farm, or suburbs and open country, or even from town to town." His plan was ambitious to say the least; he predicted that state funding for footpaths would be repaid by mass use, yet he surely knew that the audience for long-distance trails was quite small in the 1920s. The footpaths he imagined would wind their way along the natural contours of the land at a slight remove from the road, offering the walker an experience of the forest without the intrusion of close auto traffic. Such a plan would help offset the "penalty on legs" issued by the automobile boom of the post–World War I era. For a car lover like McFarland, this was a way to find a happy medium.[14]

His work on zoning boards encouraged him to apply emerging planning principles to country highways. "Public orderliness" in cities could be applied readily to the sparsely populated areas beyond. Like all other attempts to sculpt the environment in a democracy, McFarland explained to Harrisburg's Rotary Club, zoning was necessary to allow experts to teach the public what was in their best interest. In order to keep the "pig out of the parlor," he sought advice from people who were committed to a new era of thoroughfares. In 1920, McFarland consulted Philip Buttrick, a forester at the Michigan Agricultural College. Buttrick, who trained in the Yale Forest School a decade earlier, published an article in *American City* that echoed McFarland's emphasis on using "enlightened public sentiment" to pressure governments into investing in road beautification. Tree-lined ways, Buttrick believed, could transform Americans' experience of movement through and between cities. McFarland pushed Buttrick to consider how the cosmetic improvement of highways fit into a more ambitious project to reengineer the nation's transportation system. At a time when car registrations were increasing and the national railroad system had lost whatever effectiveness it once had, McFarland predicted that roads would become the crucial element of American social and commercial life.[15]

McFarland's other medium for outdoor promotion was the place at which he spent the most time when not in Harrisburg: the mountaintop borough of Eagles Mere. The history of the site as a resort began in the 1880s, when the land—a lake surrounded by dense forests—was purchased by a group of bankers and industrialists known as the Eagles Mere Syndicate. They built hotels, a bathing beach, docks, streets, and private cottages around the lake while marketing the site vigorously as a destination. McFarland arrived on the scene in the summer of 1897 and traveled there most summers after that.

The Chautauqua movement passed through Eagles Mere at the turn of the century, spurring further growth. The years around 1910 marked the time of greatest change, when the syndicate demolished the structures around the beach and moved them farther into the woods, to make an afternoon near the lake seem like time in deep wilderness. McFarland cited "world-traveled observers" who praised the lake's border of "unspoiled primeval forest growth." Eventually, McFarland hoped, the sunbather would see water, trees, and glimpses of a "dainty little tea-house toward the west, and just the tips of the other necessary buildings." While the rest of the state had devolved into "ugliness and wickedness by the sins of neglect and of greed," the resort was engineered to recede into the past.[16]

He gushed over the three road approaches to his mountain perch, viewing them as a key part of the aesthetic experience of a stay. From the east, travelers enjoyed a trip "reminiscent of Alpine journeys." From the west, the trip from Williamsport featured the idyllic valley of Muncy Creek and a road surrounded by native flora. The approach from the northwest presented the motorist with mountain views "not excelled anywhere in America." He gestured toward advertising the resort to the fifteen million people he located within an overnight trip, yet he certainly did not want crowds to ascend to the mountaintop. Eagles Mere was a 1,000-acre site that comfortably fit several hundred people at the height of the summer season. McFarland knew that a self-selection process would apply; only certain types of people read his articles or attended his talks, and only some of them were likely to visit. Who would appreciate Eagles Mere? People who were "not in love with noise and smoke and roar and racket." This was elite signaling at its best, conveyed to people who mostly lived in noisy, smoky cities and understood a temporary escape from them as a matter of taste. Whereas the masses had their play-worlds on the coast or in the cities, discerning men and women took to the trees. As McFarland tried his hand at landscape architecture, zoning, botany, and boosterism between 1900 and 1925, he honed a voice to make people listen.[17]

Cars and Cameras

McFarland wanted people to mimic the aesthetic choices he made when he went to the woods. His reach for expertise relied on his audience's faith that he knew what he was doing, especially when it came to operating in the wild. Cars and cameras became crucial to teaching people how to see well outdoors. The machine in the garden did not seem the least bit incongruous to him. He pitched his command of technology as a general command of aesthetic ambience on the roads and in the woods of northern Pennsylvania. Both popular technologies rewarded an attention to detail, which McFarland hinted was the foundation of a way to be in the wild outdoors. His extraordinary focus, he showed, had been trained over a lifetime of travel, and cars and cameras offered tangible proof of his experience.[18]

McFarland was an avid motorist, keeping meticulous records of his operating costs and distances traveled. He represented the first generation of popular motorists, people who learned that driving required technical know-how and statistical precision. He liked to display his auto knowledge, telling a federal official in 1924 that he would save the government five and three-tenths cents per mile if he took a train to Washington instead of driving. This kind of obsessive attention to detail carried into his presentation of nature. His talks were filled with references to exact mileages and driving directions. He believed that central Pennsylvania's "scenic supremacy" came from its diverse collection of river valleys and mountain ridges. The valleys, gaps, passes, and headlands made the state unique, and it was from cars that most people would experience them. By 1920, as he reached his sixties, he was content to view nature from a distance, with a chauffeur to drive him around the state and a camera next to him. This seems the epitome of the systematized, "motorized recapturing of nature by the city-dweller," yet the logic of the daring man behind the wheel remained. The roads between Harrisburg and Eagles Mere might have been improved by the turn of the century, but there was still enough adventure on those paths to mark the drive to the mountains as a gendered domain of mastery.[19]

McFarland epitomized the male touring photographer who was mythologized by camera manufacturers in the early part of the century. These were the wealthy professionals who worked relentlessly in daily life and then broke away in fits of leisure. Eastman Kodak Company specifically targeted men "who own cars and have money in chunks." Just as a camera in the hands of a motorist was meant to keep memories of trips alive, McFarland hoped to serve as a surrogate for people who needed direction in where to go and what to see. The combination of car and camera made both the journeys and the images produced during them the prize of the privileged few. Eastman Kodak's 1910 catalog of portable cameras featured fourteen pictures of people in automobiles, and in all of them men drove the cars while women, if present, rode as passengers. The company certainly used the fantasy of men eternally behind the wheel to encourage purchases. Men presumably controlled the post-trip narration of the photographs as well. Eastman Kodak stressed the parallels between operating a car in the wild and learning the photographic process. In some of the promotional photos, men conferred while fixing a tire or navigated their machines through difficult stream crossings. The catalog's writer opened with the promise of the accelerated sublime, rooted in manly control:

> In front of you, a long white ribbon of road. Behind you, a white cloud of dust. On either side, fields, mountains, a river, a valley—the country passing by. Beneath your feet, an engine purring and gurgling, the hum of exhaust droning a low note of comfort. As the throttle creeps forward and the spark slowly advances, the hum rises an octave to the middle register; it sings of the pleasures of swift motion, the joy of the bouncing springs and the exhilaration of the soft air on your face. And then, as the engine picks up, the song skies to the upper register, higher and higher, until as the air meets your face in a wayward rush that beats at the eyes and all but pulls the breath from your body, it becomes a single screaming note.

This portrayal of physical sensation checked by sensible control featured prominently in McFarland's thinking about technology in the outdoors.[20]

He believed in the lasting power of scenic mobility. He narrated specific trips along roads that lingered in his mind, such was their emotional effect. In speeches and columns, he tried to convey his memories of special roads, like the "easy mountain road" beside railroad tracks that cut through viburnum "particularly characteristic" of an area. The experience of driving along a road through rhododendrons, pines, and hemlocks that "belonged there" struck him as a profound education. It was a tactile type of learning that made everything *feel* as if it were in its right place. The visual harmony of the moving scene had to be evident to his audience, but relatively few people would have been able to identify indigenous plants. That left experts like McFarland to make judgments on the right and wrong place for particular flora, and he was quick to declare only certain trees and shrubs as "worthwhile." A kind of drive-through nature study could be had, as the speed afforded by modern cars and roads harmonized with the spectacle of engineered wildness. He trained viewers in seeing more than they ever expected on the way to their destinations.[21]

McFarland considered often how natural scenes opened themselves up to viewers. Although he believed in the power of photographs to drive interest in beautiful places, he knew, too, that they could not capture the experience of being *in* a landscape. Popular images of the Alps and the Rockies, for instance, had convinced northeasterners that those mountain ranges defined majestic scenery. Not so fast, McFarland cautioned: "the impression of majestic height relates mostly to the position from which one sees the hills and mountains." The Alps or Rockies might be high, but travelers who wanted to appreciate them found themselves too embedded in the landscape to perceive that elevation. Pennsylvanian mountains, on the other hand, were situated in ways that placed the majesty of the landscape on an ever-approaching horizon for the motorist. "There are points in Pennsylvania," he noted, "looking from which the eye may rest upon true Alpine conditions, lacking only in summer the snow-covered summits." When men pulled over at a wayside to let their passengers enjoy the view, they mimicked the production of scenery that McFarland modeled. Roadside vistas were gifts that cars gave to drivers and that drivers gave to their fellow travelers. Swift motion along narrow roads, too, created sensory experiences that passengers felt as much as saw. Cars, writes one historian of motoring, "gave back the foreground that had been lost by the train." Even so, cameras could do only so much to capture the experience of trees flying by and the wind in one's hair.[22]

The evolution of camera technology at the turn of the century meant that photographers with even meager experience could create more images in conditions that had previously made photography an ordeal. For those photographers who were not professionals, yet were also not rank amateurs, new vistas opened up. Photography was not yet ubiquitous in the early twentieth century; cameras were still expensive, and developing images was a task that required specialist skill or money. However, the new wave of cameras that emerged around the turn of the century allowed people like McFarland to produce more images with less concern about cost and effort. In his twenties and thirties, McFarland joined the swelling ranks of American amateurs who gravitated first to the smaller cameras with dry gelatin plates of the 1880s and later to the more portable offerings of Eastman Kodak. Writers in *Outing* magazine praised the "beautifully light apparatus" for affording greater access to the outdoors, but they also criticized 90 percent of amateur snapshots as "simply worthless rubbish" because the photographers knew little. McFarland built his photographic know-how slowly to counteract the haste that he saw in too many fellow enthusiasts. His advice marked him as what John Stilgoe might label a "popular" photographer who knew the language of "serious" photographers. He was conversant with many techniques and understood how to use them to build relationships with viewers, but he did not rely on image making for his income or delve deeply into the chemistry of developing negatives. He referred to himself in 1909 as a "camera fiend," and in the same year *Hampton's Magazine* listed photography as his leading hobby.[23]

By the turn of the century, McFarland *thought* through his cameras. He illustrated the view of photography as a "promiscuous way of seeing," with the mobile technology propelling users to seek out and briefly fixate on sights that they might otherwise ignore. In this mode, journeys become shaped by the obligations introduced by the camera—there are sights that simply must be captured. McFarland pressured himself to deliver scenes to his audience of prospective travelers. His approach to image making echoed the pioneering landscape photographers of a generation before who had used pictures of western mountains to entice the public and index the outdoors. William Henry Jackson's emphasis on capturing Yellowstone scenes that seemingly "demanded" to be photographed—and his recognition that he was the first person to have captured them in that way—was similar to the motivation that McFarland found along the roads and in the woods.[24]

McFarland was a devotee of the stereopticon, a projection device used in illustrated lectures. In 1898 he discussed plans to buy a $250 stereopticon for a Harrisburg Sunday school which would be used for public lectures during the week. Images could educate, he told Mira Lloyd Dock, and his goal was to "keep on increasing the information and intelligence of our people by the use of the lantern." From that point on, he trained himself to create lantern slides. His widely traveled lectures from the 1910s and 1920s rotated through hundreds of his Eagles Mere slides alone. The slide shows were meant to test the limits of photography. When natural scenes were the focus of a photographer's work, there was always the threat that they would underwhelm the viewer. In terms of size, color, flatness, and degree of detail, prints struggled against limitations that negated the deep, expansive sensations one felt within a forest. The experience of nature was elusive; many who wrote about the outdoors at that time noted that their sensations were difficult to translate for those not present. The naturalist John Muir faced this when he wrote about the mountains of the American West. He strained to convert the sublimity of his experiences into the beauty that magazine and newspaper readers seemed to want. Camera manufacturers made this difficulty part of their sales strategy and tried to convince consumers that pictures could become their *personal* memories, pleasant to others, perhaps, but never truly shared.[25]

Slide shows could retrieve some of the outdoor spectacle through enlargement, coloring, and, of course, expert narration. There was an art to these shows, as well as a commercial aspect. Lecture bureaus in cities across the United States rented slides and scripts to consumers who wanted to put up shows for guests in their own homes. An excellent illustrated lecture aimed to transport the audience to the site, a feat that could be hard to accomplish owing to the hardware at the venues in which McFarland spoke. *Photo-Era* editors spoke directly to people like McFarland in 1911 when they offered a litany of complaints from the viewpoint of the lecture audience: cracked slides created prism effects, poorly focused stereopticons made viewers squint, slides that were too big jammed in the carriage, unevenly lit screens produced unwanted shadows, and fingerprints on slides ruined the illusion of being in a landscape.[26]

McFarland was encouraged in his presentation efforts by his Eagles Mere neighbor William Simon, a chemist from Baltimore. By the time of his death in 1916, Simon had mastered the technical side of color photography and was recognized as one of the leading American practitioners of the innovative Lumière process. Using special plates coated in potato starch, the Lumière "autochome" produced images that were impressive renditions of landscape scenes. Simon introduced the technique to the Engineers Society of Pennsylvania in 1909 as a seismic shift for photography. He was remembered by the *National Cyclopaedia of American Biography* as the only man of his time who had photographed a rainbow accurately. Ads for McFarland's printing company boasted that his staff had been the first Americans to produce autochromes in November 1907 and that the results were "simply astonishing." He slowly found that they were of variable use in lectures. Although McFarland called autochromes "perfect color memoranda" in 1909, after a decade of using them he knew that the intensity of light needed to project the color images was hard to come by.[27]

When McFarland summarized his expert vision in the woods as advice for novices, he presented two related principles. First, he argued that true understanding came when viewers controlled their ambitions. They should not try to take in too much at once, he counseled, but limit their view to discrete elements in their turn. How did amateur woods walkers know what to look at or what to take a picture of? Their focus should stay on specific items—the "jewels of nature" as McFarland presented them over the decades. Visitors to the woods should attend to a lone hemlock casting its shadow on a hill or a single bluebell working its way out of the underbrush. Even visitors at overlooks must concentrate on one aspect of the scene. They should focus, McFarland advised, on the foreground at the expense of the background. When taking photographs, this meant "subduing the importance" of the background with lenses that produced a pleasant haze surrounding the object. This generally went against the advice of professionals, who recommended that photographers allow elements of the natural background to "assert themselves," instead of making them soft, unfocused canvases. Not for McFarland—when taking pictures or simply walking, he meant to maintain control. A stroll through the woods brought with it a strong sense of duty.[28]

Second, he voiced one of the most calculated explanations of the picturesque mode of viewing landscape in his accompanying rule. When instructing how to take a good photo of a tree, McFarland urged readers to treat everything in the scene that was not the tree itself as a "landscape accessory." Scenes could be composed in a photograph or assembled in one's mind by mixing and matching these accessories, walking around to include some in the frame and exclude others. Everything was a surface to McFarland, and so the difference between good images and bad images was not determined by how they commented on nature, but how they reflected beauty. Thinking of nature as a series of landscape accessories was symptomatic of a "masculinist" tradition in geographic and environmental thought which emphasized indexing and altering the outdoors. In this tradition, nature existed for human benefit, and reworking it to get one's desired result was the most appropriate form of engagement with it. In addition to patience, what the woods walker needed was "intelligent, and not arbitrary or didactic art training." If travelers allowed someone like McFarland to teach them about visual forms, they could manipulate landscape accessories at will. If we hold McFarland to his own rules about framing the natural world—following along as he tried to create lasting pictures instead of what he dismissed as "mere photographs"—we can glimpse expertise in the making. Each of the following images built his repertoire of experiences offered to his audiences of "militant citizens." Each told people what to expect and what was expected of them.[29]

Many of McFarland's photos showed scenic views from roadways, with the road as a key part of the composition. These images encouraged audience members to imagine themselves on a motor tour and simulated his assumed stance of a city dweller. An image from August 1916 captured this style (fig. 1.1). Taken five miles north of Eagles Mere on a dirt road above Loyalsock Creek, the autochrome showed three men touring the highlands with an au-

1.1. J. Horace McFarland, "Loyalsock Road." Manuscript Group 85, J. Horace McFarland Papers. Courtesy of Pennsylvania Historical and Museum Commission, Pennsylvania State Archives.

tomobile. One of the men sat in the car, while the other two had gotten out to look at the town of Forksville in the distance. One of the standing men held his hands to his face, suggesting that he was looking at the scene through viewing glasses. The town sat in the creek valley, and three overlapping mountains led the viewer's gaze into the distance.

Conventional picturesque landscape scenes placed an observer at a vista in the immediate foreground, allowing the viewer of the image to share in the perspective of the person at the site. McFarland moved far enough away from the trio that the photograph was as much about them embedded in the landscape as it was about the view that they took in. The situation he sought was this encounter between attentive tourists and the pleasant country. Locating the car in the middle distance allowed McFarland to highlight the vegetation lining the road, as well as the trees and fields in the valley below. This image would have worked well as a part of his scenic preservation lectures; it encouraged viewers to imagine not only the scenery of such a drive but also the thrill of the journey. McFarland rated the Loyalsock corridor's views as some of the best in the nation.

The photo from the road outside Forksville performed further alterations to picturesque standards. Art critics and naturalists alike had traditionally treated the picturesque as a means through which women engaged with nature. In a hierarchy of sensibilities and artistic styles, photographic tastemakers considered the interpretation of nature as pleasant and orderly as less intellectually demanding than treatments that highlighted the natural world's indifferent power, overwhelming size, and routine violence. The "higher" levels of outdoor perception, especially versions of the natural sublime, were the domain of men. When McFarland took such a picture, then, he literally illustrated his attempt to get more Pennsylvanians thinking about the prettiness of their state as an asset. This included a professional class of city-dwelling men. The men above Forksville illustrated a movement that fascinated McFarland: shifting back and forth from the wild abandon of cars to the solemnity of the roadside. A. H. Beardsley, the editor of *Photo-Era*'s "Crucible" department for savvy technicians, observed in late 1919, "It is astonishing to note the number of amateur photographers who fail to use automobile-trips to photographic advantage." Beardsley shared with McFarland a consumer's mentality, through which any drive became a chance to "obtain much valuable picture-material." He warned readers that if they were passengers, and not behind the wheel, they would contend with the driver's zeal to complete the trip as quickly as possible. McFarland recognized "picture-material" when he saw it. He subdued the background of the image by making the men, the car, and the brush surrounding them the most defined part of the scene; everything beyond them was softened.[30]

A second McFarland slide, when placed alongside the view above Forksville, highlights the gender conventions at play in roadside viewing. McFarland took this image on the same motor tour in August 1916, several miles to the southeast of Eagles Mere (fig. 1.2). The photograph showed a view of the Muncy Creek valley toward the southwest, with North Mountain cutting in from the left of the frame and a series of rolling hills and ridges receding into a white haze. This second autochrome relied on a fair amount of "printing in" clouds during development to give the sky an attractive depth. The site, known locally as "Fiester's View," was named for the Fiester family, who lived in the farmhouse shown in the middle distance. To the right of the house, in the shade of a stand of trees, sat an automobile with two almost imperceptible male figures standing by it. Closer to the camera, along the dirt

1.2. J. Horace McFarland, "Road from Laporte." Manuscript Group 85, J. Horace McFarland Papers. Courtesy of Pennsylvania Historical and Museum Commission, Pennsylvania State Archives.

road that dropped down the hill toward Sonestown, another man stood and looked toward the view. Apart from the panorama itself, the other focal point of the photograph was the trio of women in the foreground, backs to the camera, surveying the scene before them. It is likely that the car that brought them there was behind McFarland, at a makeshift parking area already worn in by motorists.[31]

The most noticeable difference between McFarland's techniques in the two images was his positioning of the surrogate viewers. The men above Forksville and the women at Fiester's View stood in for the audience members who viewed the images during McFarland's illustrated presentations. The women's status in the image diverged from the men's in terms of their relative inaction. McFarland clearly posed the women in a way that was not evident in the first image. They stood several yards away from the road, in a spot that gained them no scenic advantage but allowed McFarland to adopt his favorite "over the shoulder" effect. They stood in the foreground, whereas the men in the previous image were far enough away from the viewer to make their activities part of the composition. Moreover, the passivity of the women appeared in subtle contrast to the Forksville men. The absence of the car from the second image contributed to this effect, naturalizing the presence of the women in a way that the previous image did not. Though their dress did not inscribe them with rural status, the women appeared as landscape accessories. Audiences were meant to imagine the men in the first photograph dismounting from the automobile, grabbing their field glasses, examining details in the landscape, and then continuing on their way. The women at Fiester's View were models for the art of looking.

With persistence and hardy tires, travelers on the roads near Forksville or Sonestown eventually came to Eagles Mere. In addition to the lake, the clean air, and the genteel company, McFarland argued that the woods surrounding the resort were the prime attraction. An image that McFarland captioned "Woods View" demonstrated his vision of nature as a source of psychic rejuvenation (fig. 1.3). He took the photo in the woods around Eagles Mere in 1907. The hemlocks surrounding the lake, he wrote the following summer, gave a "good forest color." In his association of trees with the spectacle that they afforded the observer, McFarland joined a long aesthetic tradition that valued the orderly and delightful visual field. Order emerged in black-and-white slides through the rendering of light and shadow. McFarland likely shot this image from the Laurel Path, a walking trail that had circled the lake since the previous decade. Turning to the side and shooting low through the trees, McFarland captured the dense underbrush of mountain laurel and ferns that covered much of the northern Appalachian highlands. Though the photo was not a traditional picturesque scene with a distant horizon, it contained a central picturesque element. McFarland framed the scene with tree trunks that formed a doorway into the forest. They gave a sense of depth to a photo that was almost entirely foreground. Viewers' eyes likely moved to the space between the trunks, imagining a trail among the brush that was not actually there.[32]

Walking trails around the lake and into the woods helped foster a sensation of timelessness. On new trails that McFarland helped build in 1909 and 1910, guests could commune with scenery without needing the "mountain legs and mountain lungs" of the "hardy man." The Arrow Paths, groomed, blazed with arrows, and posted with regular signs, opened up the woods to women and anyone else who sought "safety, convenience, and beauty." Mc-

1.3. J. Horace McFarland, "Woods View." Manuscript Group 85, J. Horace McFarland Papers. Courtesy of Pennsylvania Historical and Museum Commission, Pennsylvania State Archives.

Farland touted the individual sites reachable via these routes, such as the unusual Eagle Rocks or Moosehead Passage, as the "best of the virgin forest." He liked to think of the resort as "shut in" by trees that had survived since the age of Columbus. The gently graded footpaths allowed even casual travelers to see these sights, but as of the 1920s few people had actually visited Eagles Mere. "Woods View" depicted the pleasant emptiness that McFarland used as part of his claim to expertise. The solitary trip into the wilderness was a powerful motif for him, despite his tendency after 1900 to visit Eagles Mere with family, friends, or his chauffeur. While the crowd splashed on the beach and dawdled on the dock, McFarland knew, the individual could break away to play at isolation. The blunt caption suggested the type of narrow focus that the walker adopted when even mid-distance views were obscured. Thinking of the landscape as a collection of accessories, McFarland found an angle that combined old trees, young trees, ferns, and broken sightlines extending into the forest. In a 1906 article in *Outlook* magazine, he called this process "hunting in the May woods."[33]

"Woods View" functioned as the "before" shot to the "after" image provided by another Eagles Mere photograph. For "Lake View," McFarland stood on the boardwalk that led to the north end of the lake and aimed his camera south, toward the shoreline (fig. 1.4). This was his own corner of Eagles Mere, where he owned a cottage ("Bide-a-Wee") and spent most of his time while on site. McFarland followed the advice of photography magazines in displacing the path in the frame, drawing the viewer's eye along it instead of allowing it to cut the scene in half. The middle ground of the image was most in focus here, with the foreground trees blurry and the distant lake indistinct in the summer light. Light and shadow mattered more than natural detail in a composition like this.[34]

The boardwalk functioned in "Lake View" in a similar way to the pathless space between the trees in "Woods View." Both beckoned the viewer forward, but the journey in "Lake View" would have offered none of the adventure that waited in the mountain forests. The boardwalk was part of what McFarland praised as sensible construction at the resort. It represented the "simple best of civilization" that made life easy—but not too easy—at Eagles Mere. A humble walkway accentuated the aesthetics of the site, whereas the "boardwalky stuff" at popular spots like Coney Island or Atlantic City obliterated their respective scenes. In this photograph, McFarland hoped to show off the boardwalk, to boast of the "New Eagles Mere" that was built after 1910. However, he did not caption the image with an eye toward the walkway, but instead stressed what the visitor saw from the walkway. Whereas "Woods View" was almost nothing but the woods, the lake hardly appeared in "Lake View." It was "subdued" as part of the background. Here was an approximation of the photographer's daily routine, walking from his house each morning to see the lake, chat to fellow residents, and figure out what to do with his day. The photo allowed McFarland to bring his audience further into his world, to recreate the experiences that made an eye trained for the outdoors.

1.4. J. Horace McFarland, "Lake View." Manuscript Group 85, J. Horace McFarland Papers. Courtesy of Pennsylvania Historical and Museum Commission, Pennsylvania State Archives.

Indeed, more often than he created views conveying pristine nature, McFarland attempted to model visitors' interaction with the mountain woodland. These were his most tourist-friendly images, ones that converted the forests of Sullivan County into the specific scenic spots that became part of the "Sullivan Highlands" in semipopular parlance. The autochrome taken in 1911 of Shanersburg View exhibited this mode of imagery (fig. 1.5). A young girl stood with her back to the camera at the edge of a dense stand of shrubs

1.5. J. Horace McFarland, "Eagles Mere." Manuscript Group 85, J. Horace McFarland Papers. Courtesy of Pennsylvania Historical and Museum Commission, Pennsylvania State Archives.

and trees that marked the slope of a hill. She looked over them, with the viewer, at the eastern horizon. In the distance sat a hazy forest punctuated by Shanersburg Run. The girl lifted a hand to her face, either shading her eyes from the day's brightness or perhaps holding field glasses. The slide contrasted the girl's impossibly white dress with varying shades of green and gray. McFarland's technical mastery of outdoor photography was on display here. He avoided the most common mistake made when photographing from vistas; by using the appropriate light filter, he achieved a pleasant contrast between the sky and foreground, and neither of them dominated the other.[35]

An image like this showed McFarland's true colors as a scenery enthusiast. His peers tended to think of the ecological whole (though not in those terms) rather than the attractive component. McFarland was in it for the individual sights that he could photograph, figuratively crossing them off of his master list. He recognized his "heresy" in using wide-angle lenses when photographing landscape, but he defended his choice as a painter might. In his photography guidebook from 1911, he was careful to distinguish his style of image making from that of other photographers who happened to be outdoors. His work, he stressed, was "decorative photography," a catchall for imagery that emphasized form and composition over the workings of the natural world. A wide-angle lens allowed him to get more background in the frame, but he never meant the background to be anything but a hazy set in front of which his subjects sat. Those foreground subjects—flowering trees, beds of laurel, girls in white dresses—could be presented to lay audiences as examples of the highlights of a drive or walk.[36]

By contrast, the men who wrote to photography magazines seemed intent on patrolling the border between serious photographs and pretty pictures. The former were made with expert ability and mechanical precision before the shutter opened, whereas the latter were doctored after the exposure had been made. One writer to *Photo-Era* complained that too many images that were celebrated by art galleries displayed a "lack of technical knowledge or power of execution by photographic means." They were pleasant, yes, but they were also "half-breed paintings." The argument was mirrored in the magazine's monthly features for two distinct sets of readers. The male-edited "Crucible" column provided a "monthly digest of facts for practical workers," such as the chemistry behind photography and mechanical processes. "The Round Robin Guild," on the other hand, was edited by women during the same period and was presented to an amateur audience assumed to have a good number of women within it. This column considered the effects of the seasons on photography, the basics of developing prints, and tips for winning photo competitions. Many writers in the national photography press followed the example of W. S. Lee, who masculinized the technical side of photography when he wrote in 1919 that "some shoot well and others shoot often, but most fail to shoot hard enough to make a sure killing."[37]

Although McFarland used his share of hunting metaphors when writing about outdoor photography, he had no qualms about embellishing images to generate the desired effect. He criticized "fake color reproductions," but he knew how to use the autochrome development process to highlight selected elements and diminish others. Of the six images studied here, "Shanersburg View" conformed most to the ideals of the picturesque. McFarland placed viewers over the shoulder of a surrogate, omitted evidence of how she arrived at the spot, and directed the image's perspective through a window framed by trees and bushes. The scene aided McFarland's praise of Eagles Mere as a tasteful family resort; thus, the picturesque elements of a tinted sky and a natural "window" onto the horizon were certainly worth stressing. The image also modeled the type of reverence before nature which he preached throughout his career of scenic preservation. Reaching this remote spot on the trail required effort from the girl along the Blue Arrow Path, and her reverent attitude suggested that she recognized the beauty before her as worth the trip. McFarland posed her in the same overt style that he used in the Fiester's View photo. Stood on a rock in the middle of nowhere, the girl was a pupil in a veritable outdoor aesthetics lesson taught by the man with the camera. His implicit claim was that the girl could receive the aesthetic message that the scene offered—presumably because of her company at the overlook. In naming the image "Shanersburg View," McFarland heralded the process by which a nondescript spot to the east of Eagles Mere became part of a constellation of branded, signed, and mapped sites in an elite tourist landscape. McFarland named many similar vistas between 1900 and 1920, using specific names to entice his peers onto trails and direct their experience once they were lured in.[38]

A final image from McFarland's collection suggested the type of immersion in the landscape which the young girl could not hope to achieve. Captioned "Primeval Forest," the image showed what it was like to be in the general vicinity of the previous photo, but down in the depths of Shanersburg Run, instead of viewing it from above (fig. 1.6). Three men stood in the bed of the creek, with bags and canteens slung around their necks and walking sticks in hand. One of them leaned against a fallen tree that served as a focusing device for McFarland. The composition was one of his standard types: rocks and branches in the foreground gave way to diagonals of terrain in the middle distance, which receded to a lighter background that was more suggested than depicted. The trees bending their way upward from the steep slope were McFarland's jewel, the "primeval forest" that he drove for hours and lugged camera equipment to capture. A working lumber camp was located within a mile of this spot, but the image portrayed an untouched wilderness. Only the presence of the men and the impression of a walking path leading from the bottom right corner of the frame interrupted the scene. The "interior" quality of the picture, with its greens, browns, and grays, reinforced McFarland's observation that there were no "pure" colors in nature. The autochrome came close to showing what the eye showed: the muted, shady atmosphere at the bottom of Shanersburg Run.[39]

1.6. J. Horace McFarland, "Primeval Forest." Manuscript Group 85, J. Horace McFarland Papers. Courtesy of Pennsylvania Historical and Museum Commission, Pennsylvania State Archives.

"Primeval Forest" showed men at ease within a landscape instead of beholding it from afar. They had gotten to the bottom of things, as it were, and for a while there were no more spectacles to appraise. The photographer made their pause in the creek bed a subject of interest, and the discrepancy in size between the men and their surroundings conveyed the scale of Eagles Mere's woods and the possibility of quiet moments in every dip. The forest did not appear large enough, however, to evoke the God-fearing emotions usually associated with the sublime. This was the "woodsy" forest, a term McFarland used to define city dwellers' estimation of a forested area that *looked right*. It played the part of stately backdrop for men's actions. They had gotten themselves there, and they would soon hike back out, reaching the comforts of the resort well before dinner was served at the Forest Inn. McFarland, too, had shouldered his equipment down the trail to this site. The logistical ordeal of photography did not dictate a purely functional style, but rather contributed to a tendency of "probing, analyzing, and active observing" in the outdoors. McFarland portrayed himself as a master of this trinity, using patient contemplation to build a visual vocabulary. He was in line with the photography columnists who encouraged woods walkers to stop and study their surroundings, to avoid the tug of the trail before them and stay in one spot for as long as it took to make an excellent image. The figures in the image certainly could have been women; McFarland wrote of taking female companions deep into the woods to see sights. But the logic of the picture worked better with men playing the roles of explorers. They had worked hard to reach a special spot, and they deserved a rest.[40]

Club-Like, Yet Democratic: McFarland and Expertise

While McFarland liked to talk often about the photographs hanging on his office walls, there was another detail of his mountain experience which he shared frequently with readers and listeners. When he stayed at Eagles Mere for days or weeks on end, he lived in a cozy cabin whose front door had a sturdy lock on it. He never used the key, however, for his time in the mountains was an idyll that would never be broken by such things as burglary. Likewise, when he used the bathhouse on Lake Eagles Mere, he never secured his valuables before launching his boat into the water or lazing on the beach. One did not need to consider such things. When he searched for terms to describe the social atmosphere at Eagles Mere, like he did in an article for *Suburban Life* in 1908, he settled on "club-like, yet democratic." The people who frequented the resort were of a type—the type with wealth and decorum.[41]

McFarland exhibited the inherent elitism of wilderness advocacy in the early twentieth century. His definitions and appreciation of wild areas were predicated on the availability of solitude, which required limiting devices to keep the masses out of the woods. The most common limiting devices, which he simply assumed, were cost and accessibility. McFarland wrote about Eagles Mere for people who could afford Eagles Mere. His carefree trust in his fellow visitors would have collapsed if the site had become truly popular. He complained in a 1928 radio address about the "ravages" of "piratical tourists." Likewise, when he promoted footpaths along country roads as a democratic device, engineered for "citizens and taxpayers" who were unable to "travel on the wings of an explosion motor," he never explained the logistics of the dream. He must have recognized that long-distance trips on foot by significant numbers of people were a fanciful vision, yet it did not stop him from speaking about new road and trail systems as if their value were indisputable. The result of this way of thinking was that if the wild *was* preserved, then it was preserved for the very type of people who relied professionally and economically on the commercial metropolis. The mountains became a getaway for white-collar urbanites.[42]

There was another aspect to the "club-like, yet democratic" label. McFarland wrote in 1908 that the woods around Eagles Mere were always there, "ready to make [him] over." The rejuvenation that he thought possible from walking through the woods was a scenic therapy for people who lived out of touch with the natural world. In his hundreds of illustrated lectures, McFarland presented the eastern highlands as a setting, an outdoor stage on which outsiders performed in prescribed ways. The setting was both natural and man-made, yet McFarland, like many outdoor popularizers, imagined that the mountains were empty. There were no locals there, no one to offer competing interpretations of wild landscapes. And so the Appalachian setting was a place for outsiders to put their commercialized lives in context by encountering scenic beauty. He imagined that people who followed his advice would find the mountains "so perfectly natural and 'woodsy.'" Whereas foresters or botanists would have scoffed at an uncritical conception of "woodsiness" (or his related notion of "sightliness"), McFarland embraced it. His favorite spots represented not simply wilderness, but rather wilderness imprinted with roads, cars, cameras, cabins, paths, and expert advice on how to use them all correctly.[43]

McFarland's direct influence on his audiences is difficult to measure. His professional reputation as an expert on scenery survived well into the 1920s, when he adopted a less demanding work schedule and devoted himself to his home garden in Harrisburg. By this point, the City Beautiful impulse had come and gone. The ACA may have paid a price for McFarland's commandeering of the spotlight. Throughout his presidency, the group had a skeleton staff and a modest membership, and the thousands of small civic groups throughout the nation found it easy to engage or disengage with the ACA at will. The result was an "ephemeral, indistinct air" to the ACA, which was not helped by the professionalization of city planning in the 1910s. Planners soon worked directly with municipal governments, cutting out the cultured, connected elites who excelled at working as middlemen. The type of physical access and mobility that McFarland relied on to project expertise became simultaneously more attainable by middle-class Americans and less compelling as a source of authority.[44]

Yet for at least a decade, McFarland was in the driver's seat of the scenic preservation movement in the United States. He knew enough about horticulture to grow roses and identify plants and trees. He knew enough about aesthetics to speak to civic groups about what was pleasant in the world. But his expertise was not in either of those fields. His expertise was the translation of those fields to a middle-class public and the fostering of a democratic feel to a decidedly club-like set of preoccupations. The perceived purity and incompleteness of remote places like Eagles Mere were the key to this claim of expertise. For McFarland, it was not a matter of William Cronon's "bipolar moral scales," which judge the tree in the wilderness more favorably than the tree in the city park. Nor was it an escape from history, for McFarland believed that what had been made by humans could be remade. In his mobile logic, Eagles Mere needed the city as much as the city needed Eagles Mere—and everyone needed him, because he had the trained eye. His audiences might put themselves in his shoes, so to speak, as he walked in the woods. Marginal places like the northern tier of Pennsylvania, when viewed from Harrisburg, were useful to experts like McFarland because of their marginality. Elite preservationists conveyed a perception of the woods as a hypothetical destination worth guarding. The people *might* pursue it because they were convinced by men like McFarland, but they might also never make the trip, secure in the knowledge that someone had already gone there for them.[45]

CHAPTER TWO

Terrestrial and Human

Benton MacKaye's Mountain Work

On September 17, 1933, the fifty-four-year-old regional planner Benton MacKaye boarded an inclined train at the base of Lookout Mountain, Tennessee, and rumbled for eight minutes to the top of the 2,000-foot peak. It was a warm Sunday afternoon, and MacKaye was but one of hundreds of people milling about in the overlook park above Chattanooga. He was not in the mood for the crowd, having spent the past two weeks riding buses and trains across the country from New Mexico, where he had been fired from his job with the Civilian Conservation Corps's Emergency Conservation Work program. The tedium of the southern plains and the sweltering heat had done him in. Lookout Mountain seemed to offer little relief. But then the rains came, and he got lucky. As fellow sightseers "squealed" and ran for cover from the storm front, MacKaye stayed at the overlook, determined to watch the rain clouds move over the valley. In a letter to his sister Hazel, MacKaye explained what happened next: "I GOT the view that I was after. In a rift between two storms there in copper colored outline through the mist was the old Appalachian Range topped by the Appalachian Trail that wound its way to the pastured hills above Gould Farm. It was the first sight of my own REGION and homeland after the dreary foreign prairie. 'Far Tennessee' became my home and country."[1]

The perceived regional link collapsed the 900 miles between Tennessee, where he now worked, and western Massachusetts, where Hazel lived. It was not the first time that MacKaye had told such a story, nor would it be the last. He did it again a year later, when he lounged near the Little River with members of the Smoky Mountains Hiking Club. He imagined once again the region stretching around him, and he wrote to Hazel about the recurring sensation. The isolated spot on the riverbank hit him with the "Primeval in full blast" until his thoughts wandered northward, "far up along the Range" to Hazel in Massachusetts and their brother in New Hampshire. His younger sister's persistent struggle with mental illness was surely on his mind; she might take solace, after all, in knowing that he thought of her. On her veranda at Gould Farm, a mental health institution in Berkshire County, she sat less than 5 miles from the crest of the Appalachian ridgeline.[2]

More than an attempt to comfort a troubled mind, MacKaye's vision of the "old Appalachian Range" was symptomatic of the mountain work that he had been doing for over a decade. He got the view that he was after because he knew where and how to look. While the "hellish tourist crowd" abandoned the overlook to the elements, he marveled at the way in which stormy weather animated the tableau. *Imagine the scene,* he urged Hazel, as the hidden form of Appalachia revealed itself to the trained observer and opened up even beyond the power of his sight. As an intellectual regional planner without a steady employer but with plenty of friends, he had spent his forties writing a book, making career plans, and reimagining American life, all the while coordinating the stitching together of an Appalachian Trail from Maine to Georgia. He had made a name for himself through his endless championing of eccentric ideas and his knack for seeing what others could not. If what he could see was not *actually* there, that only proved that he had work to do.

MacKaye's focus on the Appalachian Mountains in the 1920s and 1930s shared with the Lookout Mountain episode the narrative of epiphany which formed the heart of his professional aspirations. Like McFarland, MacKaye believed that the mountains could be a form of salvation for people who lived in cities and had become disconnected to the natural world. But unlike McFarland's emphasis on aesthetics and "woodsiness," MacKaye lobbied for a practical, even mechanical, understanding of highland landscapes. He saw more at play than surface features and urged that the processes undergirding the Appalachian forests must be comprehended and wielded by city dwellers in order to secure an encouraging future. He wanted the population of the United States which lived to the east of the Appalachians to *use* the mountains to save themselves. His was a form of literal exploitation that would unleash the latent power he saw in the topography and hydrology of the mountain range by reserving the high route for specific types of recreation and study. To convince others of that power, MacKaye devoted himself to a campaign of "revelation" which set him apart from his gang of planners and landscape architects. He argued for his own relevance by stressing the role of public relations in the professional pursuits of shaping landscape to alter human behavior—and shaping human behavior to alter the landscape.[3]

The Work

MacKaye was born in Connecticut in 1879 into a family that a recent writer has called "quirky and charismatic." He spent some of his childhood in New York City, but his family moved with the ups and downs of his father's career as a playwright, theatrical producer, and impresario. From the age of nine, his family home was in Shirley Center, Massachusetts, a town that came to symbolize for him the best qualities of colonial villages. He spent many a day in his youth tramping around the modest hills and dense woods of northern Massachusetts, something that he referenced in his public talks about the importance of spending time outdoors. MacKaye studied geology at Harvard as an undergraduate and returned as a graduate student in the school's new forestry curriculum. After teaching forestry in his twenties, he spent the 1910s employed by the federal government, first in the Forest Service and then in the Department of Labor.[4]

MacKaye's work took a turn after World War I, when he collaborated with a group of regional planning advocates based in and around New York City. Among those who would be most influential to his thinking about Appalachia as a region were Clarence Stein, an architect and future leader of the American city planning movement; Charles Whitaker, the editor of the *Journal of American Institute of Architects* (*JAIA*); and Lewis Mumford, a writer and versatile social critic. The group coalesced as the Committee on Community Planning of the American Institute of Architects and then, in 1923, formed the Regional Planning Association of America (RPAA). Although he was never quite a "committee man" like McFarland, MacKaye took advantage of a wide and productive network of professional acquaintances. It was with the support of the RPAA that MacKaye developed the Appalachian Trail (AT) as an experiment in the directed experience of nature.[5]

As his biographer, Larry Anderson, noted, MacKaye was already thinking of ways to create linear systems of outdoor recreation by 1916. He wrote about state forests in New York as a potential national recreation resource if connected into a chain. It was also in 1916 that MacKaye first referred to the idea of a footpath along the Appalachian backbone. He introduced the idea to specialist readers in 1921 and 1922, in journals of the American Institute of Architects and the Appalachian Mountain Club. In those publications, produced for groups that counseled their members to view the world as a series of problems to be solved, MacKaye presented the AT as a solution to social and infrastructural snags in the eastern United States—and as a worthwhile project for those of "genus engineer." The Appalachian highlands struck MacKaye as an exceptional region because of its "underdeveloped" status and proximity to the industrial zones of the Eastern Seaboard. MacKaye explained in 1924 that the mountain belt "penetrates the most populous portion of the United States" yet still retained the "purest form of original American Anglo-Saxon stock" and traditions of "colonial America and of Elizabethan England rather than of modern New York or Paris."[6]

The mountain ridges were close enough to population centers to allow city dwellers to enjoy their solitude, but they were thus close enough to feel the effects of metropolitan growth as well. Time was of the essence, MacKaye cautioned: seize control of the mountain ridges now and regulate them to stave off commercial development. As the "barrier of barriers," the Appalachian Mountains could be used to stop industrial sprawl. The AT would channel outdoor activities for a massive population of easterners who could drive to designated locations, walk away from their cars, and spend time camping, hiking, and enjoying the outdoors. His concept assumed that the AT would be populated once specific activities emerged to engage people in its use. The AT itself was the vehicle for a type of being in the woods which MacKaye believed was key to the transformation that he wanted to impose on life in the eastern United States.[7]

During the winter of 1922, after the idea had been formally introduced in print, he spent weeks poring over nearly one hundred topographic maps from the United States Geological Survey. He started charting the route from its southern end, aided in this effort by Jonesborough, Tennessee, resident Paul Fink, who drafted a route on topographic maps for the length of North Carolina. In February 1922 MacKaye studied Fink's route and notes, trying to match his knowledge of the northern woods with what he thought must be the experience of the southern trail. He worked in layers, completing a general topographic "highlights" map first, before moving to a map showing possible trail routes with their main rail connections. The scope of the project quickly became apparent.[8]

As the project moved forward, he courted outdoor clubs to provide building and maintenance support, turning them into the work crews that translated ideas into infrastructure. Along with the Civilian Conservation Corps in the 1930s, hiking clubs provided the local knowledge and grunt work for scouting, clearing, and blazing trails. MacKaye's project succeeded only because club members fostered the trail in their backyards in the decade and a half following his 1921 proposal. Hiking clubs also propelled the scenic preservation and planning movements in more subtle ways. Hiking clubs shared members with clubs devoted to photography, birding, wildflowers, philosophical debate, and pollution control. Allies gained in one group produced converts in others. MacKaye collaborated with trail clubs in order to gain initial control over vast stretches of the AT. By the spring of 1922, he reported to Stein, he had "six strategic connections with trail clubs." The first official stretch of the AT opened in 1923, established by the New York–New Jersey Trail Conference. Trail clubs scattered along the Appalachian Mountains were eager to get the official go-ahead to start trail work. Fink wrote to MacKaye in late 1924 to announce the development of the Smoky Mountains Hiking Club in Knoxville. This was just the kind of organization that MacKaye had wanted to see south of Washington, DC. A year later, he had also recruited the students in outing clubs from Dartmouth, Williams, Smith, and Wellesley Colleges and the Massachusetts Institute of Technology. Slowly, the AT gained the infrastructure to match the vision.[9]

While he was building his team, MacKaye came to see the necessity of a "super group" to make official judgments about the AT's course, distribute work to local clubs, and establish official signage. The process was complex and collaborative; before the "super group" could set the official location of the AT, it would have to consult with groups that had the best knowledge of potential paths. Those groups would not act on their inclinations, though, until they received the go-ahead from above. MacKaye predicted that this back-and-forth would take more than a year, making mid-1924 his preferred date for the start of blazing. In the meantime, he wrote Stein, local groups would have to scout their areas in detail and establish alternate routes that might become necessary.[10]

In order to turn interest into actual use, MacKaye knew that access to the trail would be paramount. The course of the Appalachian range brought it close to some large population centers while remaining moderately far from others. So although residents of New York might have a relatively short trip to make by car or rail to get to the AT, residents of Richmond, Raleigh, or Columbia had to expend more effort to get access. He encouraged people to think of the AT in their vicinity as an extension of "local circuits that will be interesting to populations of towns and cities." Boston was an early concern, but his knowledge of upland Massachusetts allowed him to imagine the possibilities. Boston residents could use the Wapack Trail to move from Mount

Watatic to Pack Monadnock Mountain in New Hampshire. The task then would be to get into the White Mountains. He did not imagine Bostonians walking their way out of the city and onto the AT, at a distance of over 100 miles. Rather, the idea of an approach satisfied his sense of the AT as a feature connected to a region. Whether AT users actually used such an approach was less important than establishing the potential to do so. By the mid-1920s, the Wapack Trail featured a camp at its halfway point. This enthused MacKaye considerably; he saw it as a microcosm of the "equipment for our whole Appalachian scheme."[11]

MacKaye wanted to see the AT from start to finish, planning a summer car trip in 1922 to scout the "strategic spots" where the trail came close to cities or crossed roadways. The trip along the entire length of the AT eventually happened in the fall of 1924, when MacKaye drove along the route with fellow regional planners. As he thought it might, however, the work of the trail got away from him. He told Stein in 1930 that he barely knew what was happening with the project. He felt connected to it in detail only when Appalachian Trail Conference (ATC) members contacted him directly with news or inquires. By that point, MacKaye had all but given up on the social engineering aspects of his original plan. In 1931, he wrote an open letter to be read at the fifth annual meeting of the ATC. His plan was still more than a hiking trail, but it had become more abstract. "The trail," he wrote, "is but the entrance to the final thing we seek—that thing eternal which we have called the primeval influence—that opposite thing from 'metropolitan' influence (the present passing phase of a hectic civilization)." He still referred to the mountains as a "barrier," but he spoke less about his original plans for timber conservation and protection from forest fires. Though he stressed the recreational aspects of the AT, he urged ATC members not to sponsor footraces on it or develop amusement parks around it. In the shift of the trail concept from an ambitious social experiment to a recreation space, geographer Ronald Foresta argues, MacKaye produced a "medium of adjustment to urbanism" for privileged city dwellers. This was their escape, a "place where they could briefly express their individualism." Still, MacKaye believed that outdoor enjoyment could lead to something much more powerful—if only people would venture out.[12]

Buzzwords

In his attempt to woo city dwellers onto the AT, MacKaye developed a rhetorical style that begged for the public's indulgence. He repeatedly coined new labels and phrases, trying to change his audiences' understanding of even the most ordinary aspects of their lives. The mechanics of his claims to expertise lay in the translation of the everyday world into these fanciful, even gimmicky concepts. Most often, he intended the labels to help laymen organize the world into two or three categories that made his far-fetched ideas seem grounded. To accept that these concepts made sense was to give in to MacKaye's arguments about what had to be done next. The buzzwords were always connected to prescribed actions. MacKaye shared this penchant for neologisms with his planning friends, most of whom believed that they needed to change the way that people thought about everyday routines. So highways became "cement railroads." Creativity became "psychological horsepower." Sometimes these inventions simply did not work. Over dinner with the park engineer William A. Welch in March 1925, MacKaye spoke about the Shenandoah section of the AT as forming an "Appalachia" to complement the seaboard "Atlantis." He had heard Stein refer to the Boston–New York–Philadelphia–Washington corridor as "Atlantis" days before at the first meeting of the ATC, and he used it here as if it were now a common way of thinking. MacKaye dropped the term quickly, though, never referring to it in print. There was a risk in being too fanciful. But more often, MacKaye used his phrases again and again, relying on repetition to make tangible what was in his mind. Five of these neologisms capture the core of MacKaye's motivations for the AT.[13]

First, MacKaye divided the geography of the eastern United States into three zones: the primeval, the rural, and the metropolitan. The Appalachian Mountains were part of the "primeval" region, farthest west, highest in altitude, and least developed in the 1920s. The primeval region was the expansive, if dwindling, forest, which provoked a feeling of welcome remoteness. The primeval stood "aloof from the ordinary world below." That aloofness was threatened by the world on the opposite end of the spectrum, the metropolitan zone, with its "invasion" of commercial structures and mind-sets. The metropolitan region was the city, growing inexorably as capital concentrated and population followed. As he told Norman Thomas of the League for Industrial Democracy in 1925, he considered "all cities at once" when he thought of American life and lobbied for a thorough redistribution of the national population away from congested areas. Between the primeval and the metropolitan sat the rural zone, composed of agricultural villages, modeled in his mind on Shirley Center. This was the "main expanse of hill and dale, of field and woodlot," featuring industry of the type and scope prevalent a century before. If the feeling created by the primeval was silent contemplation, then the rural evoked a feeling of community—of belonging. MacKaye observed that the "suburbanizing of the colonial village" had already started in earnest and would be followed by the "Coney-Island-izing" of the primeval mountains.[14]

Next, MacKaye used the label "entanglement" to diagnose the problem that emanated from the metropolitan zone. The idea cast the urban experience as an "industrial wilderness" and a "penetration" designed by "real estate men." Entanglement was the quaint and the local suffocated by the generic. Entanglement was rural authenticity leached away by the city. Too many Americans equated happiness with consumerism and never considered fighting entanglement. Instead of the community involvement and democratic participation of "original America," which he believed could still be found in small towns in the rural belt, citizens of the twentieth century preferred commercial distractions. He vowed not to "stand supinely by while the life force and promise of a soil's experience [were] trampled by the stampede of metropolitanism." The term "entanglement" connoted complexity, insatiable demands, and a lack of design.[15]

MacKaye championed a process of disentanglement which would promote "social efficiency" by directing industrial profit toward "human welfare." In order to get people thinking beyond the routines of their entangled lives—to shake them into demanding more—MacKaye preached visual simplicity and a retreat from commercial ideals. What he needed was to steer city folk into the wild. The AT fit squarely into the hope for natural space that was regulated but not developed in a commercial sense. The AT was a lack of development; though managed and scrupulously planned, it would stand as a meaningful blank space on the landscape. With blazes and an improved footpath, the AT promised to reorganize the logistics of life in the eastern United States. The "open spaces on the mountain crests" would form the counterinvasion, stopping metropolitan growth in its tracks and training city dwellers to go back to their communities and rein in commercial culture. This was magical thinking, relying on a transcendent experience that would thoroughly reorganize AT users' lives.[16]

Third, MacKaye juxtaposed "civilization" with "culture" in order to draw Manichean lines between cities and towns. Lest any of his readers aspire to a "civilized" state of affairs, he was quick to drag the term through the dirt. Civilization was "the period of stagnation which precedes the end" and a "cobwebbing in the winter of a human cycle." MacKaye used the term repeatedly as the epitome of all that was crass and impersonal. He thought that the urban populace lacked a communal imagination; Americans could no longer tell themselves stories about how they fit into a landscape and how that harmony of place produced national greatness. Humble towns in the rural belt, on the other hand, fostered "culture," a force of creativity and human growth. It was culture that would take root once urbanites returned from the Appalachian ridgeline. They would learn to think in terms of history, place, artistic expression, and reciprocity.[17]

Lewis Mumford advised MacKaye in March 1927 to stop using "civilization" as a derogatory term, as he had in a *Christian Science Monitor* article two months before. Thinking negatively about civilization, noted Mumford, ran counter to readers' "associations and prejudices." His friends knew that he had trouble playing nicely with the idea of cities. In his endorsement of MacKaye for employment with the Tennessee Valley Authority in 1933, Stein listed one of MacKaye's shortcomings: his "outlook and interest [were] largely confined to the open country and the small town." But if people must live in cities, MacKaye figured, then experiencing regular contact with the wild outdoors renewed their ability to animate the landscape with fantasies. "Unless an innate culture can express itself, it must die for lack of breath," he wrote in 1932. He would teach people how to breathe deeply and use their imaginations.[18]

Fourth, MacKaye described the everyday disposition of people within states of civilization and culture via the phrases "the toiling trend" and "the living trend." For MacKaye, those toward the bottom of the American social ladder simply spent too much time working for little spiritual reward. The "toiling trend" had grown in prominence since the late nineteenth century, closing city dwellers' minds to the wonders of the surrounding world. They had largely abandoned the "living trend," a state of being open to sensation, appreciating the world and its natural authenticity. Those who worked long hours in harried conditions had neither time nor energy to live as anything more than animals. The fact that American life was increasingly oriented toward toil could be seen in the sprawl of cities; population density rose as people packed together to live near their workplaces. For MacKaye, curbing the growth of the industrial city through planning would lead to greater industrial productivity. Increased productivity would lead (perhaps optimistically) to increased leisure time. The labor writer Laurence Todd encouraged MacKaye, claiming that his ideas would find a warm reception among those "from places where one would rather not live."[19]

As MacKaye perceived it, the "living trend" could reemerge on the backs of a generation of primarily automobile-based explorers who experienced an opening up of space and demanded novelty. MacKaye thought in linear terms because open spaces would best serve their functions as "levees" and "dams" if they were long enough to absorb the push and shove of the metropolis. When a Forest Service district chief defined wilderness in 1921 as a stretch of land big enough to contain a two-week backpacking trip, he, too, thought of wilderness as something that was both primeval and long. This horizontal wanderlust was complemented in some people by what MacKaye called a "vertical" wanderlust. People attuned to this drive sought out the "universal tongue" of evolutionary natural history. They knew how to live, in a way that many people had never learned.[20]

Finally, MacKaye presented the "neo-recreation/decreation" duo as the logical conclusion to his primeval/metropolitan, culture/civilization, and toiling/living pairs. He became increasingly conscious about the public image of regional planning as the 1920s progressed. Part of the appeal of using the pseudo-scientific term "neo-recreation" for outdoor enjoyment was that it would distinguish MacKaye and his cohort from "a lot of goo-goo stuff," well-intentioned but ill-considered schemes for outdoor experience. He wanted to see more national parks, but only if they were regulated in ways that made them "offsets" to commercial America and not "adjuncts." Neo-recreation was the activity that would occur on the AT, the leisurely explorations of hikers, campers, and students of the outdoors. Mumford predicted that MacKaye would "whoop for joy" when he read the work of Leberecht Migge, a German landscape architect who argued in the 1920s for community gardens, nature preserves, planned government management of natural resources, and voluntary resettlement programs. This ambitious combination of ideas fit MacKaye's brew well. He contrasted neo-recreation with "decreation," a "destructive relaxation" that offered only vicarious thrills and passive escapes.[21]

All of these novel ways of thinking about the American state of affairs came together in 1928 in MacKaye's career-defining book, *The New Exploration*. An accessible work that was alternately poetic and technical, *The New Exploration* told a story that unfolded slowly. He joked that his theme was so large, so varied, that it had to appear "between the lines." The book called for a "new indigenous America," an authentic way of living that departed from the "barbed-wire entanglement" of urban modernity. He wanted something new, an engineered combination of "simplification and a good rural economy and an active provincial life and a balanced environment and a harmonious set of vital activities." Anderson argues that MacKaye's devotion to "camp ethics" distinguished him from his fellow foresters and conservationists. It also formed the core of his forward-looking program. The new ways of living he imagined would be a result of the temporary communities rebuilt in the primeval setting and the ongoing activity on a protected landscape. If he were going to be someone to whom others listened, it would come through the peopling of the AT.[22]

Perhaps MacKaye's true achievement in *The New Exploration* was in public relations. His friends and family applauded him for bringing together an environmentalist argument that allowed lay readers to comprehend the regional planning ethos. His brother James confided that the book made a philosophy out of an "aggregate of fragments." The forest engineer Raphael Zon wrote from Minnesota to praise the book's "crystallization" of planners' scattered, "vague gropings." He was a popularizer, then, even if the book never quite became a best seller. Through his sense of humor, his liberal use of metaphor, and his refusal to approach the world around him in conventional ways, MacKaye created a palpable sense of a hidden world opened up to the reader. His was expertise through revelation.[23]

Revelation

MacKaye's aim was to turn something that everyone thought they understood into something that they realized they had never quite grasped deeply enough. This maneuver made his insight invaluable, for it allowed people to address a problem that they did not even recognize. And solutions lay up in the little-known, unnoticed Appalachian range. Because he wrote and spoke in a loose, image-filled style, one might forgive his audience for thinking that he merely spun tall tales and waxed poetic. But there was a scientific foundation to his musings which encouraged his audience to take him seriously. His chosen style was a combination of "rendering technical" the natural mountains and positing that something transcendent could be found within them. When he stood on Lookout Mountain above Chattanooga in 1933, he felt the blend of mechanical and spiritual wonder that he believed good regional planning conjured. It had happened to him before, four years earlier in Vermont. He was in Rutland to speak at the winter 1929 conference of the Green Mountain Club (GMC). He praised his hosts for building a "new mode of life" on the Long Trail in the past twenty years. Its string of shelters and hiking guides made time in the woods possible for people of varying levels of commitment. MacKaye joined a group of GMC hikers as they climbed a snowy Mount Carmel, and he felt the bliss of the experience for days after. He wrote in glowing terms to GMC president Mortimer Proctor to encourage him to keep the organization focused on the work it had pursued. The Long Trail stood for the "pursuit of nature's secrets," and, he told Proctor, "you send folks away with the lingering taste on their senses of having gained a kingdom."[24]

Gaining a kingdom meant comprehending the scale of the primeval through the up-close experience of the trail and decoding the scenes offered at high spots. Absorbing detail and translating foreign languages—expert advice made this possible for the average person. A year after his *JAIA* article appeared, MacKaye considered how popular the trail would have to be to get off the ground. Up to that point, a "certain crowd of young technicians" had contributed their talents to the idea, but he believed that "amateurs" must take part in the project for it to succeed. He spent years, then, taking the message to both specialist groups and public gatherings. At a speech before the Greenfield, Massachusetts, planning committee in October 1927, MacKaye laid it out directly: planners and the people needed each other in order to get valuable work done. The planner was "the eyes and the headpiece," sitting on top of the "body" of the people. He told the crowd that "the people as a whole unguided by expert advice and suggestion is a body without a head. We must have the alliance of the two—a 'body with a head'—if we are going to solve the enigmas of a civilization growing more complex every day." By locating the complexity in a half century of commercial growth, and not in his own explanations of what it all meant, MacKaye obscured the rhetorical strategy at play in regional planning. He asserted the value of both his vision and his efforts to broadcast it.[25]

MacKaye found himself in a strategic position within the regional planning movement. Mumford confided in him in late 1924 that the rest of his cohort was skilled at technical planning but had no aptitude for visualizing new communities "on their social and civic side." MacKaye could provide the movement with "motive power" to get people out of their habits and into new ways of living; he could get the body and the head working in unison. MacKaye placed regional planners alongside statesmen and dramatists as "visualizers," experts who spoke to the masses from positions of great insight. The regional planner used maps, charts, and, crucially, "the medium of scientific statement" to offer the public a telescopic vision of "not only the actuality of the past but the potentiality of the future." In *The New Explora-*

tion, he renamed these visualizers "revealers," and he added local historians, painters, amateur photographers, and "general appreciators" to the list. To show people how their lives could be different, planners should help coordinate the visualization of the past, present, and future with exhibits and performances. "Conquest first needs comprehension," he wrote in November 1922; "here is the new angle—the new quest—to plot and chart the tangle and thereby plan its conquest."[26]

In short, MacKaye sensed that much of his efforts should be devoted to pure promotion. "Can prophecy be dramatized as well as retrospect?" MacKaye asked his brother Percy, a playwright and poet. Yes, he believed it could, and regional planners should thus set themselves upon "equipping regions for future human living." The revealer had an obligation to reach the public, for the public had no other way to learn about entanglement. He thought that he and his peers were doing nothing less than reengineering American life. The Harvard forestry program might provide the "medium of scientific statement," but evidence of time spent in the mountains was necessary, too, for impressing upon people the accuracy of his perceptions. He told the city planner Edward Hartman that an indigenous setting could be created only when planners had cultivated an "indigenous psychology." This meant imagining the "outward social activity and living" of people in the ideal setting that was the typical stuff of planners' drafts, charts, and models. The leap from models to practice was where MacKaye could prove himself; he was equally at home behind a podium and in the woods. That is why "outdoor culture" was so crucial to regional planning, because regional planning as MacKaye cast it was building a setting that was "in harmony" with practice. Hiking, camping, sightseeing, "the introspection, the strain and stir within"—all of these amounted to a "fresh mode of living" that might wean people off of "hot dogs and jazz bands."[27]

While he sorted out the logistics of the AT, MacKaye kept one eye on the PR angle. Even before the *JAIA* article was published, MacKaye enlisted colleagues and acquaintances in New York and Boston to push media exposure for the AT. He turned to Stanley Bowmar of the Denoyer Geppert Company, which made maps, globes, models, and filmstrips for schools. In addition to his strategic location in the profession of education, MacKaye told Stein, Bowmar "knew the publicity game." Percy put him in contact with the Harvard historian Albert Bushnell Hart, whom he thought could help the project in terms of trailside history. Stein counseled MacKaye to move on the "New York crowd," bringing into the loop Raymond Torrey of the *New York*

Evening Post; J. A. Allis, vice president of the New York state bank; Isaiah Bowman, director of the American Geographical Society; New York state education commissioner John Finley; and John MacCracken, president of Lafayette College. Torrey, in particular, became indispensable to the publicity campaign. Allen Chamberlain of the ATC promised to give the AT a detailed write-up in the *Boston Transcript* and distribute reprints of his *JAIA* article to like-minded people in Boston. This fit within MacKaye's plan to increase the size of his professional community to the extent that apathy and legal issues could be overcome for every section of the AT. Unlike McFarland, he did not believe that the public could be whipped into a frenzy to support conservation programs; people had to be educated gradually and gently. Revealers had to make people think that conservation schemes or planning projects were their own ideas.[28]

Ronald Foresta notes that the people most influential in the building of the trail—the local leaders who took control of their sections—were "educated, successful, mostly young, professionals, whose vocations and activities for the trail were largely separate." Unlike MacKaye, the body of practitioners looked more for a balancing change to their lives (combining the city and the woods) than a revolutionary shift in metropolitan existence. These were people compelled by what MacKaye called the "primal desire of man" to wander and explore. "The desire is linear and not round, tangential and not circular," he wrote in 1929, "It is to proceed from one scene and sphere into another, to have at our service the full view of the continent even if we use but a particle." He attributed a specific type of wanderlust to modern Americans, which required expert cultivation.[29]

Who did he imagine these people would be? Interest in the outdoors was a must, of course, but MacKaye also labeled his ideal practitioners as "liberals at least." In addition to general outdoors enthusiasts, MacKaye imagined that members of birding and botanical clubs would be enticed. He estimated that half of the people who would make a trip to a primeval area in New England were interested in nature as a "pleasant background." This was McFarland's audience, people whom MacKaye called "frolickers" and described as "gay young pent-ups." On either end of frolickers, in terms of serious contemplation of the natural world, were campers ("people to whom the mental feel of primal wilderness is as tangible an asset as the physical feel of sun and air") and motorists (the type of person who "thoroughly enjoys the natural landscape as it comes—the views of field and wood and distant mountain"). He worried that too much publicity around parks was directed at frolickers and motorists. Frolickers had little aesthetic discernment to guide them; they responded merely to popularity and what they were told to do in guidebooks. In terms of views, they sought out "not the best but the one best known." Frolickers wanted a parking lot at the top of a mountain; they would "rush to the top, cry 'Whoopee!' at the view, and then drive down again." As historian Paul Sutter has suggested, the early decades of the century were crucial for the development of an "intellectual and cultural groundwork for a mass embrace of outdoor recreation." Those whom MacKaye labeled "motorists" might have had a superficial attachment to the primeval landscape, but even their presence on the roads of the Appalachian highlands signaled a transformation in recreation. These were the people whom Sutter described as enmeshed in a cycle of ever-expanding horizons. They used their vehicles to access the world beyond cities and towns, seeking an authentic, noncommercial experience. As more of them did the same, however, they found that they needed to go farther afield to capture that elusive authenticity. MacKaye considered motorists "serious-minded" and ready to be guided by people who seemed to be in the know.[30]

Like MacKaye, Arthur Comey, secretary of the New England Trail Conference (NETC), saw the physical trail as one part of a system that made the practices of outdoor culture possible. The system was composed of the trail itself, identified water sources, communal cook sites, blazes, signs, maps, guidebooks, shelters, "kept camps," and distributed lists of commercial food sources close to the path. If all of this could be found within a day's car trip of people's homes, it would form the outer edge of their living region. MacKaye imagined the AT in Massachusetts and New Hampshire as "the week-end breathing ground for Boston." He paid close attention to the aesthetic and kinesthetic experience of landscape. In other words, he thought constantly about what an individual would see, hear, and feel while participating in the outdoor culture of the AT. He wanted people to "cope with nature directly."[31]

For all of his emphasis on long stays in the woods, MacKaye realized that average hikers would seek the most impressive vistas as a reward. During his scouting expeditions in the Northeast in 1922, he targeted high spots and the views they offered as the main attractions of the AT. His deliberations with Paul Fink over the southern terminus focused on balancing spectacular views with accessibility. Fink reported in early 1924 that all plans he had drafted for an ideal end in Tennessee led to the middle of nowhere, with little hope of getting people on and off the trail. If the AT were left "up in the air," as Fink described it, it would be a promotional disadvantage. He voted for

Chattanooga's Lookout Mountain as the end point, but MacKaye feared that the 40 miles of rolling plains between that peak and the Appalachian chain would kill the experience for users. The AT's course thus shifted to northern Georgia, to account for both the "physical and psychological angle."[32]

Under MacKaye's hand, promotional work—the revealers' craft—happened largely without photography. He was part of a generation of environmentally conscious people who viewed cameras as strategically useful but spiritually empty. He doubted photographs' revelatory power. They stripped viewers of the ability to see how things might be different. Regional planners, he told an NETC crowd in 1927, gave a "picture of possibilities," which relied in part on fantasy. Unless manipulated, photos of mountains and trails could appeal only to aesthetic sensibilities. Fink agreed and was careful to explain to MacKaye when he sent scouting photographs that he had disregarded their scenic qualities. "Press agenting" the Smokies, Fink stressed, was not his goal. Instead, he wanted the images to impart the hiker's experience moving through the woods. Photographs served utilitarian functions for MacKaye. His instructions to AT scouts in 1922 included taking pictures as a way to index a trail section's resources, in addition to making maps and collecting supporting literature. MacKaye preferred sketches and diagrams to photos because they allowed him to make arguments about what was not visible from a vista. His approach epitomized what geographer D. W. Meinig called thinking of landscape as a system. People with this view see landscape as a "transect of intricate, pulsating networks, flows, interactions, an immense input-output matrix." This viewpoint lends itself to diagrams and illustrated models because it is optimistic about the ability of people to understand the land. The education that a hiker received in the woods was an "expansion of the knowledge (and the feeling) of primeval history" for MacKaye. In order for people to read the "record of the eternal on the out-of-doors horizon," they had to do more than look at photographs.[33]

How could he get the people to see what he saw? He asked Stein to go over his plans carefully to see whether he agreed that the materials formed a "story which to my mind . . . reads like a fascinating romance." In the 1921 article that presented the AT to the planning community, MacKaye used an elaborate narrative device to convey his point about the highland barrier. He asked the reader to imagine a giant walking along the Appalachian crest line, surveying all he could see in every direction. The giant's-eye view allowed him to mythologize the mountains, removing them from the world of mere mortals. It also allowed him to stylize the experience of walking the trail. A hiker along the crest line would very rarely see panoramic views, owing to the thick trees covering the mountaintops. Mark Luccarelli reads this as a strategic use of the landscape architecture notion of the prospect, which was a view meant to inspire active contemplation in the viewer. Sutter notes that the giant was not there to view scenery, but instead itemized resources in the surrounding land.[34]

Raymond Torrey shared MacKaye's bias toward diagrams, sketches, and topographical maps. Torrey had explained in 1923 that his *New York Walk Book* featured landscape drawings by Robert Dickinson rather than photographs because cameras tended to flatten and deaden scenes that should be conveyed with more life. An artist with a pen could accurately capture the most salient feature of mountainous landscape: depth. Sketches revealed slopes "as steep as they feel," whereas photographs condensed topography. For MacKaye, the goal was to "bring out a point," not to impress with beauty. His disdain for photography extended even to portraits. When Torrey asked him for a picture of himself, "preferably in outdoor garb," to accompany a *New York Evening Post* article on the AT, MacKaye balked. He had no pictures to offer and was loathe to "go out and strike an attitude and get someone here to snapshot" him. It went against his very being, he pleaded, "to make a 'still-movie' of myself."[35]

What might have seemed like a bold choice—to conduct more than a decade's worth of AT promotion without the use of scenic photos—was all part of the plan. MacKaye celebrated a type of looking that was encouraged by the Scottish planner Sir Patrick Geddes. Like Geddes before him, he presented standing at a mountain vista as the confluence of surveying a landscape and narrating a local history. The static view in the distance became an animated, evolving scene that featured the overlapping and interacting of humans and the natural world. He believed that vistas did matter, but only when they allowed viewers to see the land arrayed like a map; that regional perspective was instructive. When he envisioned the AT as an experience for the user, he worked within a type of virtual space to rotate it and zoom in and out on it. Almost from the very start, he imagined the AT not just as a linear phenomenon, but as part of a larger system of forces and flows that moved from the heights down to the sea. The geographical cross section was his favorite device to envision this system, and its power stemmed from its focus on the whole. A trip on Vermont's Long Trail in the summer of 1922 convinced him that cross sections were the way to reveal the AT concept to the public. Unlike a panorama from the crest, a cross section made a view where none existed, allowed the viewer to "trace [the forest's] inner constituents and workings," and let the viewer "absorb" the primeval influence. A cross section inserted a study of natural resources into the visual field of the AT. MacKaye initially hoped to do a cross-sectional analysis for the whole trail, but he never got that far. Cross sections did appear in his published works, and they delivered on his hope in 1922 to present something that "amateurs" could follow.[36]

When he wrote a decade's progress report in *Scientific Monthly* in 1932, he used panoramas and cross sections as his main tools of "unraveling the story told directly by the great outdoors." The forest that visitors saw before them was an integrated part of overlapping phases of geological, botanical, and human development. Charts of processes accompanied the text, which modeled the type of trail guide that MacKaye proposed, but never wrote, for the AT. Visitors would take the guide with them as they explored interconnected trails. He knew that hiking clubs were making long, linear treks with plenty of ups and downs, but common users would expect something more manageable. He asked Frank Lutz, an entomologist and curator at the American Museum of Natural History, in March 1929 for advice on developing small sections of the trail into guided walks. Experiencing the outdoors correctly required a conscious effort to interpret that which evaded close scrutiny. The pleasant and picturesque could become, under guided observation, the basis for a "potential influence or form of psychological energy analogous to waterpower." For hikers to read the outdoors like a sensory script, they had to be trained to think of it in terms of narrative. The primeval woods offered the "greatest story in the world—the greatest adventure story, the greatest mystery story, the greatest love story." MacKaye was not kidding when he described his sense of the primeval woods as a drama. He thought of regions in terms of dramatic presentations. He wrote to H. G. Merriam of the University of Montana in 1932 about his thinking on regionalism since the publication of *The New Exploration*. Sensory impressions and scenes from his childhood formed the core of his region concept. For example, the American West (which he admitted did not appear in the book) was for him a mixture of Owen Wister's 1902 ranching novel, *The Virginian*, and the spectacles he saw when the Buffalo Bill Wild West show hit New York in the 1880s. These sources made for him a "regional articulation" that he wanted to mirror for Appalachia. In a draft of the letter to Merriam he explained that his sense of the East came from a "New England barnyard, the northwoods logging camp, and the Kentucky mountains."[37]

His concept for nature trails developed by 1932 into a proposed series of guidebooks about many AT segments, written in the style of Lutz's own nature trail guides. A regional trail system with publications devoted to its use, MacKaye reasoned, would serve the interests of museums, bus companies, and publishers (and bring him much-needed income in the process). He thought of the book as part of the system. He wrote to Clarence Stein that the project would begin formally in a local context, with a thorough ecological, topographical, geological, and historical survey of a short trail. Next came the development of a transportation system to get urbanites to the trail, followed by the creation of the pocket-size book. He suggested that the project could be called the "Nature Guide Service"; Stein countered with "Appalachian Outdoor Museum." Both labels captured the idea that walkers with books in hand would be directed by a narrator who had walked the very paths and selected which environmental aspects to highlight.[38]

In mid-1932, MacKaye threw himself into writing the guides. The trick was to write about the outdoors in a way that corresponded to the self-reflexive, fluid experience that he wanted walkers to enjoy. He explained it to his sister Hazel in a way that captured the multiple perspectives he hoped each individual could adopt. With a guidebook in hand, hikers would "see, trace, and read." Seeing meant using panoramas to get a general orientation to the landscape and indexing the plant and animal life therein; these had to be "carefully studied views," not just pretty ones. Tracing meant using linear progress along the trail to chronicle two cycles: the movement of soil and water from mountain to lowland and the short-term growth of forests. This required several elements, including a labeled cross section to show how crests and oceans were connected by erosion, a "botanic-zoologic garden" to reveal "who's who in the forest," and a path through cutover woods to exhibit the forest life cycle. Finally, reading meant visiting another vista to interpret evidence of geological and botanical evolution. As he devised it, nature trail visitors could choose to follow any of these emphases in order to tailor their experience to their interests or the natural features accentuated at a given time of the year. The collection of local books would build off of the master volume, dubbed an "Appalachian Primer of Outdoor Nature."[39]

The project proposal that he completed by November 1932 pitched the endeavor as a means of keeping the American middle class connected to nature despite the Great Depression. MacKaye was explicit about the class context of nature trails in the cover letter to his proposal, noting that the financial collapse had created two groups: the desperately poor and the relatively fortunate who sought "cheap and wholesome luxuries." Time on a trail was such a luxury, he hoped. A trip through the wild would become a compelling option for people with disposable income if they could connect it to "some fresh appropriate mental interest" that distinguished it from the "mundane worn out attractions of the street." To a great extent, his project sounded much like the *New York Walk Book* from a decade before. The guide would not convert people to the value of the outdoors but would add a new layer of meaning to outdoor experiences. He assumed that readers would be people already invested in "nature pursuits." Yet MacKaye wanted it to be the start of something grander. He was adamant that it would not be "just one more nature book"; it was an inherent part of a new way of living.[40]

A sketch that he made for the 1932 draft, showing "Appalachian America" from Tennessee to New Hampshire, illustrated the point (fig. 2.1). The perspective of the sketch placed the viewer above the Atlantic Ocean, scanning the region from the east. MacKaye included major geographic zones, like the coastal plain and piedmont of Virginia and North Carolina and the Allegheny and New England Plateaus, with rivers cutting across them to the ocean. The centerpiece of the sketch was the dark spine of the Appalachian Mountains, dwarfing the skyscraper smudges that indicated cities along the Eastern Seaboard. *All of this* was Appalachian America, including the unseen neighborhoods and commercial corridors in which readers spent the majority of their lives. People living in those tiny cities had to account for the role of the mountains in the forces and flows that made their daily routines possible. They could study that presence closely on the ground and through the "Primer." This was a tough task; MacKaye confided in his sister that the book would not be profitable. "Were it tutti-frutti it might," he wrote, but what he had in mind was "mental and not salivac." A book devoted to explaining the practices of outdoor culture on the AT catered to "enlightened outdoor trampers of the eastern USA" only, and he guessed that they were one in a thousand. Still, the proposal for a guidebook that he would never write brimmed with optimism.[41]

Trail promotion continued well into the fourteen years that it took to form a continuous path along the Appalachians. MacKaye spent much time in the twenties and thirties selling the idea of the AT and the activities it would foster. Some within the trail movement forgot the second part and focused instead on the physical trail as the end in itself. Fink reminded attendees of the 1935 Appalachian Trail Conference that the completion of construction meant only that everyone involved could turn their attention to the "task of selling the Trail to the rest of the world, just as it was sold to us in the first

2.1. Benton MacKaye, Appalachian America Sketch. MacKaye Family Papers, ML-5. Courtesy of Dartmouth College Library.

place." Fink noted that MacKaye always kept his eyes on trail use as the goal, and he encouraged others to do the same. After all, he observed, "there are hundreds of thousands of people living in sight of the skyline to whom 'Appalachian Trail' is yet only a name, nothing more."[42]

Conclusion

MacKaye did not rest when his initial coordinating efforts for the AT ebbed, though he turned to stints of employment that involved more concrete problem-solving. He worked for the Tennessee Valley Authority in the 1930s, cofounded the Wilderness Society in 1935, and was employed by the Rural Electrification Administration in the 1940s, conducting projects of his own making as an "industrial engineer." He died in 1975 at the age of ninety-six, after three decades of active retirement. Scholars cite his legacy as threefold: he made "simple but large concepts" resonate with people who had the wherewithal to put them into action, he modeled a type of distributed effort that regional planners adopted in later projects, and he sought a hybrid understanding of environment which broke down false dichotomies between wilderness and civilization.[43]

In terms of MacKaye's professional cohort, the publicity work for the AT was an attempt to bring the regional planning movement out of its closed-door tendencies. For him, environment was "part geography and part folks"; he declared, "it is terrestrial and it is human." It might be too much to say that MacKaye "brought the folks back in," but he certainly tried to instigate a circulation of like-minded people in and out of the mountains in the name of ambitious planning goals. A product of a family with clear talents in the theatrical world, MacKaye could think of complex processes occurring under the soil and in the trees as pageants that would charm visitors. He could imagine that people who wanted to drive into the mountains might appreciate precise descriptions of their feelings as they stood at an overlook or sat in the shade of dense timber. But his focus usually fell on what people could do for the mountain forests; he showed sporadic interest in the rejuvenating or life-affirming aspects of outdoor experiences. He thought less of the therapeutic experiences that McFarland sold. He needed people to get into the highlands so that the highlands were already spoken for by the time the developers and capitalists came knocking.[44]

The marketing program, as he devised it, demanded a constant push to define the ephemeral and to capture the elusive. His work can be thought of as an attempt to isolate moments of intense feelings and harness them, as if passing them over the blades of a turbine. What exactly did that magical entity, the indigenous, *feel* like? In yet another letter to Hazel, MacKaye wrote of the sensation of a mountain farm in eastern Tennessee in the summer of 1934; this was "THE most indigenous place yet," and it left an aftertaste. When he visited the farm with Knoxville friends, he felt something complex and deep. The feeling combined the look of architecture ("log house and barn"), the sounds of domesticated animals ("mooing cattle and bleating sheep . . . and cow bell tinkling in the distance"), a sense of the hard-won order kept in and around the house and in the backyard garden, a spectacular backdrop ("the timbered mountain overhead"), and a fantasy of the past ("ancestors who had sat around the same big stone fire place"). That was the mix of emotions and sensations to which he wanted to expose more people, in order to get them hooked on a feeling—the feeling of place, used as a mobilizing force for grand schemes.[45]

The notion that a place could be "the most indigenous" epitomized MacKaye's striving to quantify and control the immaterial. It was an impulse that led him briefly to consider discarding the "regional planner" label in favor of "regional engineer." The latter seemed more action oriented, more in line with the practices he wanted to devise for actual people, rather than the schemes his colleagues devised for impersonal populations. An engineer studied lines of force and adhered to exact specifications in order to problem-solve in individual cases. Engineers rendered technical, converting the complex world into component parts. A "regional" engineer did so for the make-believe qualities of places and people's experiences of them. His concern over the name and image of regional planning was a symptom of his varied, diverse training. As Sutter has noted, MacKaye worked in an "intellectual polyculture," in which boundaries between politics, science, and art held little meaning. He did not so much popularize a regional planning ethos as discover a way in which it could be broadcast via a specific project. The AT never became anything more than a fringe phenomenon in his lifetime, but it gained distinction among planners as an ambitious idea executed in a careful way.[46]

MacKaye's achievement in the 1920s and early 1930s was to elevate a specific version of the authentic to the top of many outdoor-minded groups' to-do list. But it bears repeating that the value of the mountains, in MacKaye's scheme, was to protect what lay below them. That is where the indigenous practices took place and that is where the next victim of commercial, urban ways lay. If he rarely used the term "authentic" to describe what he was after, it was perhaps in keeping with his choice of a spirited vocabulary. The authentic was the "indigenous," and it was something that could be found in the small towns of the East, not up on the slopes of Pack Monadnock or along the ridges of the Smokies. Many small towns like Shirley Center had already changed from MacKaye's postcard view of them; they had been stretched by the centrifugal force of automobiles and connected via radio to near-immediate information and far-flung entertainment. But he wanted their residents to seize control of future growth, to think about the lessons of flow and force which they could find on high. Regional planners "discovered" and "revealed," but they did not invent. The invention he left up to others.[47]

CHAPTER THREE

The Stern Grip of Circumstance

Hiking the Smoky Mountains, 1920–1940

When the Japanese-born George Masa died of a sudden illness in his home of Asheville, North Carolina, in 1933, a local newspaper reporter revealed that he left behind photographic equipment and maps. Any regular readers of local newspapers might have been shocked by Masa's death, but few would have been surprised by the artifacts remaining in his office and apartment. For almost two decades, Masa, born Masahara Iizuka, had worked steadily to build a name for himself as the operator of a commercial photography studio and a seemingly tireless recorder of the mountains that sat to the west of Asheville, along the border of North Carolina and Tennessee. His images illustrated tourist brochures. He dabbled in producing scenic postcards. He was a quintessential picture man for whom no job was too small and no slope was too steep. He made the maps and images that lured others into the highlands.[1]

If maps generate an aura of authority by suggesting that someone had been on the ground, measuring and surveying, in the very spots depicted, then Masa's maps earned their conceit. Masa spent years trudging along the rough trails that wound from high passes to even higher ridges, up weed-choked streambeds to forested saddles, and down from the highest peaks in the southern Appalachians to coves and hollows that sat in the shadow of mountains for most hours of the day. He pushed and dragged a homemade cyclometer, crafted from a bicycle tire and an odometer, to measure each twist and turn. His maps offered hikers a measure of certainty that would otherwise be hard to come by in the jumbled terrain of the Smoky Mountains. The guidebooks that accompanied the maps promised that after so many miles and so many turns at so many landmarks, travelers would find delightful overlooks, intersections with other carefully described trails, or perhaps paths back to their cars.[2]

Masa's maps made another claim as well, one that was crucial to his membership in a regional network of hiking enthusiasts, photo enthusiasts, and enthusiasm enthusiasts. The products of his work in the wilds of eastern North Carolina argued that white-collar Americans could do extraordinary things through careful organization, the disposable leisure time offered by professional occupations, and a willingness to suffer physically in the name of a specific goal. Like his maps, Masa's photos claimed that the possibilities for ordinary people were much greater than they might assume. The products of his mountain wandering could assist those people in imagining what the mountains held or in finding those very spots, if they were willing to work hard enough.

The efforts of people like George Masa—people who hiked, photographed, mapped, charted, narrated, and chatted about the southern mountains—framed the Smokies as a region in which individuals could test themselves and others in the outdoors. Masa was part of a movement that produced two hiking clubs that turned the practice of walking through mountain forests into a matter of expertise. In 1923 and 1924, men and women of the regional middle class situated on either side of the mountain chain created the Carolina Mountain Club (CMC) and the Smoky Mountains Hiking Club (SMHC) to organize their recreation in the woods. The leadership of both clubs soon portrayed hiking in the Smokies as a two-tiered pursuit, with casual, fun-seeking hikers enjoying pleasant days out and serious, trial-seeking hikers enjoying grueling ordeals. Within the confines of the clubs—and through their outreach to mass leisure audiences—the leaders claimed expertise by demonstrating that they had learned exacting lessons out on the trail. The privilege of suffering on the slopes of the Smokies was a special type of knowing that club leaders feared would become more difficult to achieve as the most remote spots of the mountains became less remote. Recreational infrastructure improvements, ushered in by regional boosters and the National Park Service (NPS), threatened to make hiking in the mountains more conventionally appealing. But there was a finite period of time in the southern Appalachians during which organizational hikers claimed a specific set of skills within their regional social circles. They touted their ability to negotiate the Smokies' obstacles in ways that few could match.

The Immensity of It All: Early Photography in the Smoky Mountains

Cameras were everywhere in the southern Appalachians in the twenties and thirties. Hikers lugged their portable models up mountains and across streams to capture views from celebrated vistas. Logging companies contracted with commercial photographers to document their widespread operations. Forest Service agents photographed trees and waterfalls to chart changes in water flow and erosion. National Park committees from North Carolina and Tennessee used photographs of serrated peaks to lobby for federal protection. Pilots flew planes rigged with cameras to help estimate the forest coverage (and thus the market value) of the most remote areas of the Smokies. Civilian Conservation Corps workers and NPS landscape architects snapped before-and-after scenes to show the scenic effects of new trails and roads. Traveling photographers offered portraits and landscape scenes to locals and tourists. When the SMHC printed an acrostic for "Smokies" in 1927, the K stood for "Kodak records."[3]

The ubiquity of cameras and the conscious drive to produce imagery were closely related to the physicality of difficult hiking. Historian Douglas Brown argues that the increasing portability of cameras made early mountaineering a reflexive activity for the climbers. Lugging cameras and thinking about photo opportunities made mountaineers perceive their actions as if removed from them. The perfect photograph of alpine beauty might be a by-product of climbers applying their skills and achieving their technical goals, but it also conditioned them to see the "abstract quality" of mountaineering. Brown's insights apply well to the hiking cohort in and around the Smokies. They fetishized the extreme effort required to do such things as climb for hours up "neck-breaking, back-bending bear trails." Their ability to make photographic records of themselves and the sights they saw at the top of those bear trails influenced their experiences. It was not just the view from and serenity at the overlook that rewarded them for their "work"; it was also the knowledge that they would have proof of it and documents that they could repeatedly study with others.[4]

George Masa walked his way into the story of southern Appalachian photography, hiking, and promotion. Along with Knoxville-based photographer Jim Thompson, Masa produced images that were both commercially appealing and useful in such technical fields as botany, cartography, and landscape engineering. He started selling postcards to Asheville tourists in 1932, a year before his death. He wrote to craft magnate and philanthropist Fred Seely that "most every tourist" demanded scenic postcards instead of 50-cent photographs. Masa reckoned that 5-cent postcards could be lucrative. The consumer-friendly sheen of these souvenirs masked the sheer strain necessary to create them over the previous fifteen years. In an interview with a local reporter, Masa stressed that he had worked in the mountains to make the "perfect pictures" that he had only dreamed about. He kept a large map of the Smokies on his office wall, using colored pins to mark the peaks that he had scaled and photographed. The journalist lingered on this map, referring to it as Masa's "textbook and the visual expression of his plans and dreams he has made for his work." Masa maintained and sold the Asheville area's most current set of hiking maps in the early 1930s. The *Asheville Times* described him as one of the "greatest experts on the North Carolina side of the Smokies in the western part of the state," one who had "tramped through them and photographed the hills and the valleys from every conceivable angle." Getting the right angles meant working his way into locations that required long walks over rough terrain along primitive paths or through utterly trailless areas. For Masa, the pictures were worth it.[5]

An example of Masa's mountain photography shows how a rather anodyne image could mask the work required to capture it. In a black-and-white image of Mount LeConte taken from the east in the late 1920s, Masa revealed a midrange shot of a peak that had not been displayed publicly before (fig. 3.1). Unlike most of his shots, this photo featured a vertical orientation. As he often did, Masa used trees in the immediate foreground to present depth. Two trees, on either side of the image, created a window through which the viewer saw the mountain in the distance. Masa took the shot from a height, from a spot near the state line north of Newfound Gap. Shooting from height meant that Mount LeConte's famous "rise" was not rendered dramatically in the image. As a mountain without a single spectacular peak, LeConte appeared as a ridge with little to mark its significance. The top third of the image was a bright sky, with little visible detail to show distinct clouds. The bottom two-thirds were the dark mass of LeConte and the valley separating it from Masa. Gradations of light played across the valley, making some points stand out brightly. It is difficult to tell the time of day in the shot; the sky was almost overexposed. Judging from the shadows of the trees in the distance, it was likely midday.

The photograph represented the effort that Masa expended to achieve shots of scenic grandeur. According to Samuel Robinson, an Asheville optometrist, Masa specialized in three types of photos: detailed images of rhododendron thickets, stark images of "serrated peaks," and hiker's-eye images from mountain vistas. The LeConte photo, lying somewhere in between the latter two, captured his behind-the-scenes practice. It was the product of a physical ordeal. By the mid-1930s, the NPS had improved enough of the

3.1. George Masa, Mount LeConte from the State Line between Mount Mingus and Newfound Gap. North Carolina Collection, Pack Memorial Public Library, Asheville, North Carolina.

LeConte approach trails to make the journey to his vantage point a matter of "simply" fighting altitude and steep grades for a few miles. Before those improvements, however, the trip was something else entirely; one had to overcome severe impediments like miles of undergrowth that grew high enough to be "overgrowth." Robinson noted that Masa endured a "great deal of hardships in making his way over rough terrain to a good viewpoint," only to wait for hours through "rain and windstorms for the right lighting conditions to obtain the picture that his artistic sense envisioned." Carrying a pack full of filters, film, a tripod, and his 8 × 10 view camera, Masa started hiking early in the morning or camped up on the state-line ridge to get such shots.[6]

The image did not reveal that slog, but it did suggest the character of movement through the Smokies. Geographer Yi-Fu Tuan argued that photographs teach viewers how to "see time 'flowing' through space," as perspective draws the eye from the "present" of the foreground to the "future" of the horizon. The viewer of Masa's image would likely realize that the country ahead was wild, but the image's point of view did not suggest that one might easily walk through the frame, down into the valley and then up the eastern slope of LeConte. The intervening trees on the sides of the frame bracketed the view as a hiker might have seen it from the trail. The effect was such that the viewer had poked his or her head through the branches to spy the view across the valley. Masa tended to frame his images with tree branches, pulling them into the frame to achieve his compositions. A critic noted toward the end of Masa's life that he succeeded where others failed in mountain landscape photography by capturing depth with this framing device. The technique had a good American pedigree, too, one that spanned painting and photography. Hudson River School artists had used it almost a century before. The masterful Yosemite photographs of Carleton Watkins frequently featured sentinel trees in the foreground, to set off his distant geological subjects. Masa's intentions were dramatic; he let the viewer in on secrets, scant seconds in time in places that were often inaccessible to all but the most determined hikers. His devotion to the shot was an extreme example of the type of patience that outdoor photographers of the early twentieth century, like J. Horace McFarland, preached to amateurs. The editors of *Outing* told readers in 1917 that the best photographers had "patience unlimited and time to waste." Masa turned the time "wasted" in the Smokies into intimate shots, conveying the illusion of being on the scene. He offered a hiker's point of view.[7]

In the buildup to national park designation, Smokies boosters disseminated mountain imagery far and wide. The Knoxville Chamber of Commerce reportedly spent as much as $300 each month on reproducing images from local photo studios. Jim Thompson became the leading photographer associated with the Tennessee marketing campaign for a national park. He and his brother had run a photography studio since the early days of the century, and in the 1920s Jim began specializing in imagery shot within the mountains. His colleagues in the booster community raved about his ability to get deep into the difficult sections of Tennessee and North Carolina to collect photographic "evidence" of what lay within. The Knoxville Chamber of Commerce official Carlos Campbell, in particular, credited Thompson's pictures with convincing federal decision makers that the national park was worth investing in. Enlarging the photos to show the scenic details of rarely viewed valleys and ridges apparently won over the Appalachian Mountain Club's Harlan Kelsey and railroad car manufacturer William Gregg.[8]

Thompson was a perfectionist, insisting that good mountain photography could not be rushed while making exposures or approached casually during the developing and printing processes. When he took part in the promotional hikes designed to impress visiting dignitaries with the Smokies' majesty, he used a trio of helpers to carry his equipment to the heights. Then, when making photographs of hikers for tourist ephemera, he insisted that too many people within the frame ruined the effect. Conveying the message that crowds of people made it to the best overlooks in the Smokies was misleading; it was better to show a single person in the frame. And they should never pose for the camera as if the photograph would be *their* souvenir. Instead, Thompson directed his subjects to turn away from him, looking at the scenes with the eventual viewer of the photograph.[9]

Thompson and Masa created expectations for people who wanted to portray the mountains for others. If they cornered the market by 1930, it was because it was a very small market. Campbell, a Smokies booster extraordinaire who made his home in Knoxville, later wrote that he had no sense of what lay to the east in the mountainous country, let alone what specific sites could be reached by determined walkers and motorists. For him, the Smokies barely existed as a concept until regional residents like Masa, Thompson, and other members of hiking clubs brought them to life in photographs, newspaper interviews, slide shows, and private correspondence. As in other "fringe" places that became part of wide mainstream tourist circuits, the Smoky Mountains' appeal benefitted from the increasing distribution of landscape scenes in newspapers, magazines, brochures, and souvenirs. The more that hitherto little-known spots of the mountains appeared in the marketplace, the more potential visitors could imagine trips and expect certain scenes to present themselves for further photography. The segmentation (even commodification) that came with such picturing paralleled what Robert Campbell argues happened along another remote spot, the Inside Passage of Alaska, three or four decades earlier. The aesthetics of kinesthetic experiences became paramount.[10]

Paul Fink admitted such an orientation when he reached impressive vistas. His usefulness to Benton MacKaye, after all, had come through days upon days spent out in the woods. Fink represented the professional, city-and-town milieu in which most Smokies wanderers lived. He and his wife, Lena, were active in the middle-class social life of Jonesborough, Tennessee, attending Sunday teas, Methodist Missionary Society gatherings, Schubert Club recitals, and meetings of the local Garden Club. Born in 1892, he was the son of a longtime Jonesborough banker. He considered hiking in the Smokies as a means of getting away from the everyday routine and a wonderful way to impress upon others the unimaginable beauty spied from the highest spots. When he pulled himself to the peak of Mount Chapman in 1919, for instance, his first reaction on seeing "one of the grandest views" of his life was to grab his Kodak camera in order to convey some fraction of the beauty to others. The conventions of scenic photography made hikers recognize when they were in front of something with special value. When he made a trip into the woods, Fink always carried on his back cameras, spare film, a set of photographic filters, and his tripod.[11]

The need to see more drove Fink to try his hand at aerial mountain photography in 1922, accompanying a friend who operated a plane. He had been looking forward to the chance to see the mountains from above, to "grasp the idea of the immensity of it all." The view from on high did not give him the holistic understanding that he sought, but it did solve a few riddles for him. Seeing the winding, twisting divide that marked the state line from above suggested why it was so easy to become lost when following what would seem like the most obvious pathway through the Smokies. Eight years later, when he tested a national park map made by the Aerotopographo Corporation, he judged it fairly useless as a hiking aid because it understated the steepness of mountain slopes. A Tennessee game warden gave guidebook writer John Faris a similar assessment, telling him to stick to the trails: "You lose your perspective up there."[12]

Aerophotography practitioners trumpeted the advances they made between 1920 and 1940, while still acknowledging the disadvantages of using their products on the ground. Military and industrial operations might benefit from photo surveys of natural resources, but hikers who depended on accurate representations of slopes and trail distances could be thrown off by the complexity of shadows. Although aerophotography companies scheduled their flights between 10:00 a.m. and 3:00 p.m. to reduce the length of shadows, they could not eliminate the effect by which mountains looked like valleys and vice versa. Trips by foot through mountainous terrain might verify the details of "aeromaps," but they could not be guided by them. Apart from his brush with the view from above, Fink stuck to the trails.[13]

Masa, Thompson, and Fink fostered their mountain photography within the region's two leading hiking clubs. But whereas Masa and Thompson were professional photographers, everyone else who encouraged and assisted their work in the 1920s and 1930s, like Fink, approached photography (and hiking) as a technical, expressive challenge. The SMHC formed in Knoxville in 1924 when Albert "Dutch" Roth and friends "realized that the people of East Tennessee knew little about the beauty of the mountains" and thus pledged to protect the highlands. Roth was a pipe fitter for the Southern Railway; most of his twenty- and thirty-something colleagues were in the early stages of white-collar careers. The first six presidents of the SMHC were a YMCA secretary, a florist, a photographer, a botanist, a lawyer, and a mail clerk. They drafted club membership rules to state that "any reputable white person" could join. Reputability and racial segregation went hand in hand in 1920s Knoxville; although the hikers escaped their lives briefly while up in the mountains, they never thought to challenge the codes of the day. Beauty remained a watchword for the club, despite the fact that many of its activities were decidedly practical. The club coordinated group hikes, trail maintenance, and educational presentations for Knoxville civic groups. Club officials prided themselves on organizational efficiency, with trail crews reporting to trail section captains, captains reporting to the chairman of the trails committee, and the chairman reporting to the board of directors. SMHC hike leaders, those who were charged with scouting trails, maintaining group morale, and planning for contingencies, were thoroughly "ordinary" people with experience in the woods. Among the hike leaders in the first five years of the SMHC were an accountant, a bank manager, a secretary, a stenographer, a manager of a furnace company, an architect, a bookkeeper, a professor, and a car salesman.[14]

The CMC formed almost simultaneously, when the southern chapter of the Appalachian Mountain Club separated from the main organization. Like its Tennessee counterpart, the CMC organized trips, issued internal hike reports, and got the word out to local residents about the joys of the mountains. Both groups combined a philosophical attachment to nature and life outside urban areas with the more mundane pursuits of trailblazing, maintenance, and photography. CMC leaders presented their club as a service to the people of western North Carolina, especially those who had never witnessed "places of rare beauty." Both groups were founded as vehicles to get more people (of a certain type) into the mountains. Leaders wanted what George Revill and Charles Watkins call "a citizenry that understood the workings of the countryside." These were people who knew how to behave, what to enjoy, and how to talk about their experiences. The organizers of the CMC, like those of the SMHC, were professionals who ran in urban circles. Leaders of the CMC in its first decade included a nurseryman, a photographer, foresters, a bookseller, and an editor/publisher. Several members were active in real estate and development by the mid-1930s. They learned about hiking and the mountains by venturing out with other members.[15]

The hiking clubs' devotion to photography stemmed from their need to generate and maintain interest, but it became a way of thinking about their activities in general. The SMHC started an annual photography contest in 1932 to encourage members to create and share their images. Jim Thompson offered a few tips, encouraging photographers to edit images; "one picture just submitted," he observed, "would rate very high if the contestant had only trimmed off some objectionable detail in the foreground." Judges rated the entries on "interest, composition, technique, and originality" and hoped that the competition would spawn "more critical observation" while on the trail. CMC members also maintained a steady interest in photographic technology and practice, at a time when Asheville housed more than a dozen professional photography studios. L. D. Rogers, chairman of the CMC's committee for photography and maps, reported in the spring of 1933 on developments in color film for everyday cameras. The club started its own photo competitions in 1934. When a camera club formed in Asheville in 1935, CMC members rushed to join and synchronize the organizations' activities. Throughout the 1930s, the photography committee of the CMC urged members to share their snapshots at meetings, as a way to improve group storytelling. The committee also solicited pictures to recruit new members and keep a club album. Before the formal competitions started, Masa requested that all CMC hikers photograph "interesting things" on their trips, resulting in many landscape shots from peaks and close-up shots of rocks or plant life.[16]

By the 1930s, then, there were plenty of individuals and firms photographing the mountains. Commercial studios established conventions for Smokies scenes associated with souvenirs. The Littlefield and Steere Company sold "Smoky Souvenir Candies," chocolates packaged in boxes with a choice of three landscape images, from cigar stands in Knoxville's major hotels. The slogan, "The views are beautiful. The candy is delicious," was typical tourism bluster, but it highlighted the distance between regional boosters' vision of what the mountains offered people and the emerging hikers' vision of what people could take from the mountains. If the hiking clubs sold their philosophy in a similar way, their box might read, "The views are beautiful. The hiking is grueling."[17]

Signs of this difficulty could be glimpsed in the advice that photographers shared with each other in the twenties and thirties. Club members warned often about dense undergrowth and fickle weather. Myrtle bushes and pine branches may have impeded perfect views from the tops of peaks, but even the most exposed vista offered little if the elements did not cooperate. All of the earliest published accounts of trips through the Smokies mentioned the clouds, wind, rain, lightning, and fog that made mountaintop observation a matter of good timing. Some authors downplayed the effects by stating that the landscape became "uncertain" in such conditions, while others warned readers that "scarcely a day was spent in the mountains [without] the distant rumblings of thunder." The Smokies guide John Oliver described the experience of reaching the top of Gregory's Bald as an attack by the elements: "The fog came down, the clouds seemed to envelop us, and the mist began to fall. We could scarcely see a stone's throw in any direction." Upon reaching a fogged-in Siler's Bald, Harvey Broome of the SMHC recalled that he "felt that we were surrounded by all the sadness and loneliness in the universe." The problems with fog and clouds were a matter of position and experience. University of Tennessee botanist S. H. Essary noted that "fleecy clouds" looked pretty only when viewed from a distance. When swathed in them, "the hiker sees nothing and sometimes gets wet in the bargain."[18]

Masa knew the difference between the experience of hiking through the mountains and the photographic scenes produced in the mountains. His trail logs were spare in their detail, recording a May 1932 trip up Grandfather Mountain in western North Carolina, for instance, with this entry:

> 0.0 Camp
>
> 0.27 road turns to left
>
> 1.43 plank road forks take R
>
> 1.67 Water—on L
>
> 2.1 plank road forks take L
>
> 2.43 shack at R spring at L
>
> 2.73 sawdust pile at R, saw mill site, leave plank road turn in R follows trail
>
> 3.4 Top—west side—then follow trail East side Top. Short distance.

His narrative of a previous four-day backpacking trip through the Smokies included exactly one description of scenery: "3:15 PM Top of Mt Chapman on Tenn. side, one of beautiful landscapes as moss covering high boulders and Natures [*sic*] hands did wonderful work." The rest of the log offered mileage, directions, and technical data useful for those who would follow in Masa's footsteps. This was but one example of the hiking clubs rendering the Smokies into an array of technical data. Masa's experience in mapping the Appalachian Trail in 1929 and 1930 had taught him how high-route paths moved across the land. His feel for the relationship among mountains, a quality that viewers noted in his Smokies photographs, was a hiker's sense of the jumbled rhythms and views atop the highest peaks.[19]

When he led SMHC hikes in the late 1920s, Dutch Roth took photographs that conveyed a little of the disarray that could be found on and off the trail. These images tested viewers' sense of orientation, displaying the Smokies as a realm with multiple depths of field. On the one hand, in his formal landscape photographs, he accentuated the mountains' monumental, if "cluttered," nature. Many of Roth's images could be read as promotional shots that portrayed the immensity of individual peaks or the wonder of multiple ranges arrayed next to each other. In his photographs of SMHC events, on the other hand, Roth displayed more limited distances, portraying the sensation of being hemmed in by the woods as one walked to the next open spot. In particular, he showed the closed-in, almost suffocating nature of hiking in the woods. His images demonstrated that a trip with companions through the Smokies was movement through a dense landscape, fog shrouded, rife with intervening tree limbs and buzzing, "hungry punkies," and with few true vantage points. Being in the Smokies was more sensuous than it appeared in most professional photography.[20]

These modes of imagery can be seen clearly in two photographs from 1927. Roth took the first from the top of the Chimneys, a pair of peaks on the Tennessee side of the divide, a few miles south of Gatlinburg (fig. 3.2). Roth positioned himself on one of the peaks and took a photo of the landscape to the east. Carlos Campbell estimated that the 2-mile hike from the main trail took three hours. The result was a collection of five layered mountain ridges, interrupting each other at wild angles while extending horizontally beyond the frame. The image captured one small section of the Smokies, yet it suggested the endless variety of the highlands. It was not as crisp as Masa's image of Mount LeConte, owing to the haze that obscures the details of the forested slopes. Nor did Roth use a framing device like Masa, making the image seem less composed and more like a glimpse than a studied gaze. Roth's caption noted that the image showed Newfound Gap before the state road had been cut, and indeed, there was no indication in the image of any man-made "improvements" on the landscape. The caption was the only orienting device in an image that otherwise depicted the mountains removed from any larger

3.2. Albert Gordon (Dutch) Roth, "Made from the Chimneys before the Road to Newfound Gap Was Cut," image no. roth0404. Roth Photograph Collection, University of Tennessee Libraries, Knoxville.

context. This was a customary Smokies scene—one that few people ever saw directly, certainly, but similar in tone to the more technically successful images that boosters soon attached to pictorial maps, newspaper features, and souvenir boxes of chocolate.

The second photograph would not sell confections (fig. 3.3). A man, likely SMHC member and future president Guy Frizzell, sat in the foreground, at the base of a rocky waterfall and brush-covered slope that overwhelmed the frame. Frizzell, a mail clerk at the University of Tennessee Experiment Station, looked directly at the camera, with his back to the water splashing all around him. The setting was Huggins Hell, a pathless area of dense rhododendron and mountain laurel which became notable within the ranks of the hiking clubs for its unforgiving difficulty. Roth again noted in his caption that the date of June 1927 meant that this was "before the Park Service took over." Even after the NPS moved in, its employees barely touched Huggins Hell, leaving it to the black bears and the rare hiker. Roth's emphasis on the NPS's entry into the Smokies as a watershed moment suggests that he used such a photograph to boast to others about the club's exploits in its early days. The image of Frizzell in the middle of Huggins Hell was meant to impress. Popular travel writer John Faris noted in 1932 that venturing into this "anoma-

3.3. Albert Gordon (Dutch) Roth, "Huggins Hell Trip before the Park Service Took Over," image no. roth0102. Roth Photograph Collection, University of Tennessee Libraries, Knoxville.

lous region" was "unnecessary and unprofitable." Hikers who ignored the advice would encounter such a frustrating combination of obstacles, injuries, wetness, and lack of views that they would wish for an immediate evacuation. The SMHC itself warned of losing sight of the companion who chose to drop into the undergrowth of Huggins Hell. "From above," the guidebook observed, "he cannot be seen, his progress being marked by a ridiculously feeble disturbance of the bushes." Although Roth and Frizzell did not make the descent solely for the purpose of creating an image, their effort to get to this spot was clear. They had slid and scraped down the mountainside that filled the frame. It may have taken the pair less than three hours to reach the spot from the trail above, but the effort would have been greater than that required for Roth's image from the Chimneys. The picture was all foreground, with only the varying size of leaves and pattern of shadows delineating the distance between Roth's lens and the surrounding forest. This was not a photograph of a place, but rather a photograph of conditions.[21]

Few hikers bothered to snap pictures within laurel slicks, lacking both the open space to maneuver a camera and the ambient light to produce a satisfactory image. Those who did take pictures in such spaces hoped to record some small aspect of experiences that could be endlessly discussed in club meetings but were rather hard to convey to the uninitiated. Author, SMHC member, and hike leader Laura Thornburgh took such photos, though they never found their way into her published works on the Smokies. While teaching summer education courses at the University of Tennessee in the 1920s, Thornburgh also wrote Smokies pieces for the national press. In these articles, she praised the "impenetrable forests, jungles of laurel and rhododendron, [and] unscalable cliffs" that she had found when guided to Siler's Bald. The time at the top had repaid her for the climb's torments, she assured her readers. Like the rest of the club hikers, she distinguished the eras before and after the work of the NPS. Before 1932, the scramble to the top of Mount LeConte was hellish, a "true test of one's mountain climbing ability." After the trail work, the climb was exciting but simple. Thornburgh wrote with a balanced tone that told her readers that they had missed out on a lot of the wild fun of the Smokies, though they likely would not have attempted such feats anyway. Her first climb up LeConte in the 1920s had been unrelenting; she and her party had "climbed over huge boulders, crawled under trees, tripped over roots, barked our shins, scratched our hands, struggled up and ever up, panting for breath."[22]

A few of her photographs from the late 1920s depicted moments of effort in the middle of nowhere. Two particular pictures that showed hikers making their way along trails gave viewers little information to figure out the entire scenarios. Instead, these were crowded close-ups, impressing upon the viewer the tangle of underbrush, narrowness of footpaths, and severity of slopes. Her intended viewers would have been fellow club members. The first image, from 1926, showed two companions from behind as they climbed up a rocky but moderate slope (fig. 3.4). Sunlight cut through the canopy in this exact spot, giving Thornburgh a spotlight on the backs of the man and woman who concentrated on the trail ahead of them. Besides their figures, the light gave viewers a glimpse of the small boulders they had to navigate, the leaves of mountain laurel bushes on either side of them, and, in the darkened background, the tangle of trees through which they would soon make their way. Without being exactly claustrophobic, Thornburgh's photo showed just how limited one's field of vision could be in southern Appalachian forests. They were thoroughly unglamorous figures, this pair turned away from the camera amid terrain that was too immediate to be scenic. The photograph was an affirmation of the process, stripped of the promise of spectacular sights.

The second photograph, taken in 1928, depicted two people on a steeper slope, shot from the side (fig. 3.5). A meager amount of sunlight again gave

3.4. Laura Thornburgh, Hikers, ca. 1926, no. IV-U-hiking-6319. Courtesy of Great Smoky Mountains National Park Archives.

Thornburgh the chance to show the details of Smokies hiking. Here, both figures turned to face the camera, though the features of the hiker in the foreground were obscured in darkness. Both hikers leaned slightly forward to climb the mountain, with the leading leg of the front figure a good foot above the other. This was a brief respite in the middle of a climb that appeared to be without the aid of a trail. Rocks, roots, weeds, and twigs were all that lay along the route. Thornburgh wrote in 1937 of a similar situation, a journey back to camp from Mount Guyot which was all "sliding and hurtling down the mountain side; slipping on the round, mossy rocks, stumbling, falling, rising only to skid and slip and fall again." The photographer was just

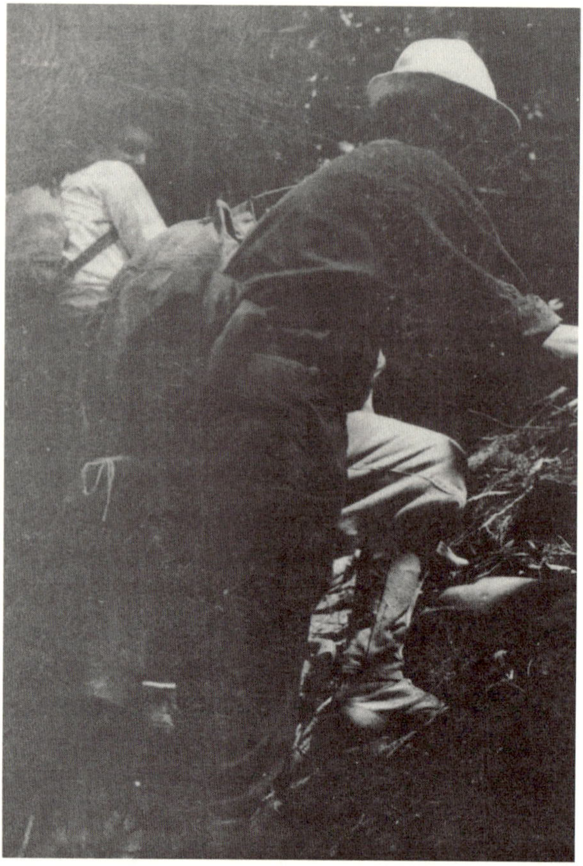

3.5. Laura Thornburgh, Hikers, ca. 1928, no. IV-U-hiking-6318. Courtesy of Great Smoky Mountains National Park Archives.

one hiker in the chain of hikers, pausing to snap the image and create distance between the lens and the group. Hiking photographs such as this were not mere representations of landscape. They were what Douglas Brown has called "an artifact of a kinetic event, the act of photographing." Thornburgh likely used such images as her own memento, although she might also have showed them to fellow club members to share memories of the trail.[23]

The photos chosen for Thornburgh's book, published under her pen name "Thornborough," were mostly images taken by others, particularly the Thompson Company, Carlos Campbell, and the Knoxville Tourist Bureau. Several of the images were her own, showing mountain residents in front of their homes and on their porches or shots of local homesteads without people. Her images were much more intimate than the others in the book, which presented the traditional promotional imagery of distant mountains and tourist facilities. Thornburgh's images appealed to a different imagined Smokies, one that inspired the NPS staff to implement cultural heritage programs in the late 1930s. But images like the trail shots had no place in promotional campaigns. They were instead part of a self-narrating project, visual records of the routines of hiking. They registered the effort needed to work one's way through the Smoky Mountains.

An Endless Amount of Trouble: Expert Hiking as Flow

The leadership of the SMHC and CMC identified with the rougher aspects of the mountains, and it was through a cultivation of that deliberate roughness that they elevated the work of people like George Masa into a set of arcane practices. Within ten years of the groups' existence, their leaders had studied the physical and psychological requirements of hiking in the Smokies enough to develop a general sense of what it took to perform well in the woods. They counseled their rank and file on how to hike optimally, in order to give themselves the best chance to want to keep hiking. When the SMHC's annual booklet boasted that the leaders had lured ninety-four hikers out into ankle-deep snow on a nasty day in early March 1930, it was a message to club members that they should continue to push themselves to deal with the challenge of mountain conditions. Beginners could slowly become adept at moving through the Smokies, and skilled hikers could earn expert status with time, effort, and focus. This was a prime example of what historian Steven Gelber calls "achievement-generated pleasure from worklike leisure." It would not always be fun, but it could usually be endured and then remembered as worthwhile. Everyone could learn how to enjoy the unpleasant, at least when discussing it weeks and months later.[24]

Harvey Broome, a Knoxville law clerk and president of the SMHC in 1932, defined his club's first two decades as a walk on the wild side:

> We have had our accidents. We carried a girl out on a slab torn from a rotting buckeye, and fished a man from the swollen Road Prong. We have waded streams in the rawness of winter; and have broken through ice with our bare feet. We have skidded around on top of LeConte at twenty below zero; and we know some of the rules for camping safely at ten below. We have climbed the snowbound cliffs of the Bunion in winter and in fog. We have fought the turbulent madness of the spring winds which can snap off great trees.

The "we" in this case was a relatively small number of hikers who participated in the club's most extreme trips. By 1930, two clear types of members had emerged: those who wanted short hikes that emphasized beauty and companionship, and those who wanted long hikes that could be spoken of afterward as an ordeal barely survived. North Carolina State Forester J. S. Holmes credited Masa's "willingness to take an endless amount of trouble" for his success in the photo business, but the sentiment applied equally to the achievements of the upstart hiking clubs. The boundary between enjoyment and suffering shifted ever upward, as club leaders urged members to learn to live with the pains of going greater distances or finding new routes.[25]

As many scholars have interpreted it, this experience depended on a tradition of staging "masculine struggle in the frontier." And, to be sure, club leaders branded areas like Huggins Hell as masculine realms, reporting in 1928 that "only eight men—and no girls" had ever ventured into that "virgin wilderness." But they seemed self-aware when using such gendered language, in a way that they never conveyed about their racial exclusion. They implied that they were in on the joke and saw the silliness inherent in white-collar professionals playing at wilderness explorers. Although women made up a minority of hiking club members, they held offices and participated in all types of organized activities. Though the longest and most arduous backpacking trips were typically male affairs, most weekend trips in the 1930s included women. Neither male nor female club members seemed to think twice about a 16-mile hike in October 1931, which they described to a local newspaper as a "chestnut hunt." Crucially, women also served as trip leaders for both the SMHC and the CMC. This meant that they participated in the narrating and scouting practices that were central to the early years of these organizations. It is tempting to read the hikers' love of the wild as a reaction against "feminized" consumption and comfort. More influential to the distinction that they attempted to erect through their mountain exploits were the markers of class. They did not value hardiness as a form of manliness as much as a pose that combined middle-class sensibilities with an acknowledged outsider status. This echoes the performances of English rambling clubs in the preceding decades. Ramblers' narratives of their "intense physical and emotional relationship" with the rugged environments of the Peak District were class-based celebrations of strenuous experiences that did not feature regularly in white-collar occupations. Organized hikers in the Smokies did not "go wild" as much as exercise their general commitment to thoroughness, technical mastery, and self-improvement in new venues. Club members—lawyers, teachers, bankers, artists, business owners, scholars, planners, and municipal officials among them—took their professional orientations with them into the woods. When Fink wrote of the trail's hardening of physiques "softened by months of office work," he meant that all of those hours in the green tunnel of a viewless hiking path were good for something after all.[26]

Early CMC trips were as much about cutting trails as they were about scenery, making the core membership a group of people dedicated to working in the woods. By 1933 the CMC had completed fifty group hikes averaging 10 miles in length. The club was just as organized as its Tennessee neighbor, routinizing the process of trip reporting with forms that recorded timing, mileage, personnel, directions, altitude, and scenic observations. In the process, hiking stories became regimented. Tales of bad decisions and twisted ankles proliferated. In 1932 club members returned twenty-seven hike reports, which the guidebook committee compiled into a manual for club use and public sale. By 1935, the CMC held a yearly competition for the "best story" submitted by trip reporters. The competition's organizers asked for great detail in these reports. Their goal was to produce aesthetically minded, adventurous narratives that created tales for the landscape. They wanted more than distances, times, and altitudes. After his guided trip through the Smokies with Paul Fink, Appalachian Trail Conference (ATC) president Myron Avery worked to get more coverage of the southern mountains for northern readers of the ATC's magazine. He contacted the CMC early in 1932 with an offer: in exchange for new subscriptions to *Appalachia*, Avery would publish CMC hike narratives. Stories became standardized.[27]

When MacKaye interacted with the SMHC's "merry crew" while living in Tennessee in 1933 and 1934, it was usually on hikes or at hike-planning gatherings. The *Knoxville Journal* reported on MacKaye's club activities, noting for readers exactly when he planned to see his first sunrise on Mount LeConte with his guides. MacKaye read to them from this book, attended banquets with them, and, within a year of meeting them, told his brother that they were "the most enlightened group of young people" he had ever seen while on the AT. But MacKaye, in his midfifties, was not part of the more strenuous outings that the club devised. He appreciated how much they thought about their trail experience, even if he lacked the time or the stamina to follow them wherever they went.[28]

The clubs' focus on process coincided with myriad efforts to cut trails through the Smokies. As congressional approval of the Great Smoky Mountains National Park became clear in the late 1920s, Fink pushed his fellow hikers to build better trails. He argued that new paths needed a "definite reason for being, as leading to a peak, a cascade, groves of big trees, or a short route from one main trail to another." To Fink, proper trails gave the hiker the opportunity to look at something impressive. He hoped, moreover, that future trails to peaks would avoid choked hollows to grant hikers exciting views on their way to the main attraction. "In addition to leading one to his destination," he urged, "there must be enough of interest along its route to justify one in following it." Landscapes created by hiking trails, then, should be participatory, unfolding over linear miles and along various points of altitude.[29]

Fink, like many hiking narrators, thought on two levels: one concerned with the hiker's experience of movement through space, and one fixed on the creation of place when the hiker stopped. Tourism scholars have emphasized the divergence between the landscape views of insiders and outsiders. Insiders, those who live on the land, often value process over product. Karoline Daugstad notes that insiders' perspective tends to be "activity-oriented," allowing scholars to consider the "embodied landscape" as constituted through human practice. Outsiders, for the most part, experience the landscape "from a certain distance and mainly through the eye." Hiking clubs tried to juggle both views, orchestrating the experience of the mountains on the way to and at the vista. Fink was in full promotion mode in 1929; new trails, he argued, should serve the needs of newcomers—people who wanted to be entertained or educated by their path through the woods. Club leaders like Broome, Campbell, and Frizzell looked for a kinesthetic expertise that came through sore feet, aching backs, and burning lungs. They did not require customary features of "interest" to push them along.[30]

The hikers who led the clubs were hell-bent on achieving a state of optimal activity which scholars have examined since the late twentieth century. Although they certainly did not have the vocabulary to indicate as such, the hikers in the Smokies were grasping for a state of "flow." Developed by psychologist Mihaly Csikszentmihalyi, the concept of flow describes the experience of a state of highly efficient effort and psychological harmony. Flow happens when people are deeply focused on the practice of a certain task, to the point that they feel a rewarding, energizing ease of progress. What distinguishes flow from a more generic notion of being "in the zone" is that flow can be experienced through deliberate organization. The club work of the Smokies hikers was crucial to this planning; the anticipation of and achievement of flow on the mountain trails was a highly structured affair. That structure allowed them to become the type of people whom Csikszentmihalyi has studied: "people who devote much energy to some activity which yields minimal rewards of a conventional sort." It allowed them to enjoy themselves not only when they were at the impressive spots from which Masa photographed but also when they were in the sticky situations that Thornburgh captured.[31]

Csikszentmihalyi argues that flow can be designed by following a five-step process. Each of these steps corresponds to a type of activity that the regional clubs pursued in the late 1920s and early 1930s. First, *one must set a goal or a series of feasible subgoals that structure the overall activity*. A group goal should be tangible enough to be conveyed to those involved, and the meaning or value of achieving the goal should be clear. When the SMHC hikers decided to complete a thru-hike of the Tennessee / North Carolina state-line route from one end of the national park to the other, they established a goal that was ambitious for 1932. To their knowledge, two groups had unsuccessfully attempted the journey, a track record that made the trip even more appealing. But they hedged their bets, announcing to the rest of the club that they would attempt only the eastern half of the journey, with Newfound Gap as the destination. That summer, the SMHC leaders conducted this "experiment" hike, a trip of more than 30 miles. Campbell claimed that, without announcing their intentions to anyone but the nine members who took part, they planned to continue past Newfound Gap, ending at Deal's Gap, on the western edge of the national park. The 72 miles in eight days became a legendary mark of daring. Leading the trip, with Dutch Roth, was Bob Burch, a twenty-five-year-old grocery salesman.[32]

When the members of the CMC decided to explore Huggins Hell in 1931 and 1932, they performed the same type of structured goal setting that served a crucial function in Csikszentmihalyi's theory: they provided a "feeling of novelty and challenge," a chance to explore the limits of individual and group abilities. Huggins Hell loomed large in the thoughts of southern Appalachian hikers. This 400-acre patch of vegetation was named after a local herdsman who was lost within it for days before following a stream to open forest. The 1931 SMHC handbook claimed that Huggins Hell was a " 'brawr'-infested, ivy-barricaded, boulder-strewn, water-soaked, viewless concentration of exasperation and torture which for the average hiker is neither interesting nor beautiful," yet was "overwhelmingly appealing to those with a spark of the devil in their blood." Six members of the SMHC had dragged themselves through it in the summer of 1926, and the tale of them wading the length of a waist-high creek in order to avoid the tangle on either side became legend. The CMC hikers wanted to create a standard operating procedure for the area. The challenge was not only the way in which the space obstructed movement but also the literal damage it inflicted. By targeting Huggins Hell, the CMC leadership set themselves up for an activity that could be perfected only through great effort.[33]

Step two in Csikszentmihalyi's process is to *adopt means of measuring progress toward the goal.* This provides the scale by which participants know whether they are performing well; it is the expectation of and achievement of high performance that produces the flow state in autotelic (intrinsically rewarding) activities. In the case of end-to-end hikers, this step meant reaching daily distance goals before camping at night and assessing whether they could continue when they reached the natural halfway point of the Newfound Gap highway. Hikers segmented their walks through the Smokies by the climbs they had to complete, such as the "heartbreaking task" of walking from Grassy Patch toward Newfound Gap in the days before the road was built. The distance was usually less daunting than the lack of reliable water sources, the reliance on "impassable" paths, and the twenty-two meals that members had to carry on their backs when they left the comfort of their cars for a week in the woods. The 1932 club schedule promised that those who joined the hike would soon grow accustomed to "pioneer simplicity and privation," as they climbed out of their occupational and consumer lives and into the "stern grip of circumstance with no choices at all." The club advertised the hike's features: "acute discomfort," hunger and thirst, frequent downpours, night after night of chopping firewood, constant vigilance for deadfall trees, and the requisite blend of "toil and pain and fear." Two years before, the SMHC annual guidebook described a two-day, 20-mile trip as a "super-hike." The national park thru-hike was of a different scale entirely.[34]

To highlight the masochism that one must possess to join such an endeavor, the SMHC scheduled a second, "regular" hike to accompany the thru-hikers for the first 2 miles of their journey. This, too, was a form of measurement. The day hikers were asked to bring a canteen and a lunch and to prepare themselves to help their colleagues carry leaden packs to the top of White Rock. Machine shop draftsman August Krahwinkle, leader of the short hike, noted, "We have only about two miles of easy trail to cover with them, while they have about a week of the roughest, toughest trail a hiker ever encounters." Thru-hikers had one climb by which to compare their efforts to that of the average club member. It was through such a juxtaposition that they received a near-immediate sense of how they were progressing.[35]

The CMC's Huggins Hell expedition crew created means of measurement by turning their general exploration of the uncharted area into such tangible tasks as identifying reliable entry and exit points. Time spent in Huggins Hell was not a good gauge of achievement, for merely being in the tangle of trees and brush was not a challenge. Moving through the area and orienting oneself to conditions that defied orientation were the true complications. The goal of exploration was less a matter of how long the hikers could withstand its obstacles than how they could start to make sense of it for others. In other words, whereas the goal of the state-line hike was to move from a specific starting point to a specific end point, the goal of the Huggins Hell activity was to learn how to accustom oneself to thoroughly unwelcoming conditions. Locating a "front door" and "back door" to the area was a sign of progress. Future explorations would pick up where the original group left off.

The third of Csikszentmihalyi's steps is to *focus intensely on the experience of the activity in progress,* itemizing the physical traits or kinesthetic skills that emerge as essential to performing beyond one's past abilities. This is the study phase, the time spent on the activity that is meant to break it down into its component parts. For the distance hikers, this could mean recognizing exactly when backpacks began to hurt their bodies after prolonged motion and adjusting their stances or packing strategies accordingly. It could mean noticing when the "rubbery legs and scorched lungs" of the first hours or days transformed into sturdy walking and easier breathing. It was, in short, about learning to be comfortable in the Smokies. The key was to find an appropriate level of outdoor comfort, constantly pushing it further away from the comforts of everyday routines. Historian Phoebe Kropp explains comfort in turn-of-the-century camping as something that helped people interpret the "corporeal experiences and social meanings of their endeavors." The social meaning of club hiking was to immerse oneself, with others, in the kinesthetics of hard walking.[36]

When Broome entered "another world" during his trips into the Smokies, it was not a paradise of scenery and inspiration, but instead a world "as real as gnats and flies and hard beds and blisters and sweat could make it." Focusing on what those elements felt like in practice and how one could mitigate them was essential. The "sloppy, malodorous black muck" that hikers had to scrape from their calves after walking along the top of Miry Ridge had been part of the Smokies hiking world since the early 1920s, when Fink first experienced it and told others about it. The SMHC hikers wrote about being covered by the sludge as an expected feature, the physical experience one gained when one made it to the top of the ridge. When Campbell edited the club's guidebooks in the late 1920s and early 1930s, he included a bibliography on hiking and mountains which he wanted club members to consult. One of the books, from which Campbell also quoted in guidebook epigraphs, was Charles Coleman Stoddard's *Shanks' Mare*. This celebration of walking by the New York–based writer and newspaper editor was published in 1924 and featured Stoddard's loving descriptions of the process of long journeys. Walking for hours on a footpath, argued Stoddard, brought about a "hypnotism of recurrence that dulls the conscious faculties, but lays bare the subconscious being to the most delicate impressions from within and without." This preoccupation with the physical sensations of hiking, what geographer Tim Edensor calls an "overwhelming awareness of the body that can dominate consciousness," is found in Csikszentmihalyi's flow-seeking figures; their focus is such that they become unconcerned about the possibility of losing control over their bodies. In hindsight, it seems to them like they had control over their environment, even when they were knee-deep in the mire.[37]

For the Huggins Hell explorers, this stage could mean learning that the crab walk deemed necessary to move well under thick rhododendron bushes relied on different muscles in the legs and arms than they typically used when walking upright. Carlos Campbell described most trails in the Smokies as "veritable tunnels under the canopy of leaves and branches" that allowed hikers to see little but plants. Huggins Hell was all plants and mountainside. Adding to the myopic nature of the experience were the packs that weighed hikers down. Fink noted that by the early 1920s, he and his fellow hikers had started using tumplines to reduce the pain of straps digging into their shoulders. These straps across their foreheads came at a cost, though; a hiker using a tumpline had to bend forward under the weight of the 50- to 60-pound pack. Looking at the sights while trying to walk ended up dumping the pack to the forest floor. Best to keep one's head down. The bulk of packs made movement through Huggins Hell as awkward as could be. The CMC crew moved forward only by simultaneously staying low, keeping one's body suspended above the ground, turning to the side to keep one's pack from snagging on shrubs, and keeping an eye on potential paths ahead. Moving a single mile could take an entire afternoon.[38]

Fink emerged from one excursion through such rough land with hands that were as "bleeding and raw as pieces of beefsteak." To those unprepared for the Smokies, a Knoxville resident warned, the mountains offered only "bruises, sore limbs, sufferings and tortures self-imposed." In addition to the typical physical rigors of backpacking, the undergrowth turned ankles and then cut legs, arms, and faces when hikers stumbled. In the spring of 1931 CMC member George Stephens completed a 35-mile trip, half of which he "had to crawl in order to force a way through the undergrowth." In the Big Creek section of the Smokies, Stephens reported, the undergrowth was so thick that his party crawled only 2 miles in one day, mostly on old bear tracks. This type of travel was an "acrobatic maneuver," with hikers slowly learning the arts of "squeezing, twisting, swinging, jumping, cooning, [and] testing each foothold." This pain and effort was the practice of Smoky Mountain landscapes, and it needed to be understood before hikers could begin to enjoy themselves.[39]

The third step in the flow process was how the hiking clubs spent much of their organizational energy. All communication to new members or prospective members contributed to this intense scrutiny of the mechanics of hiking, because club leaders assumed that novices were in it for the sights. They would need to convince them that there was something more to experiencing the Smokies. Le Roy Jeffers, an international champion of all things mountainous, wrote in 1922 that southern peaks offered views of "a dreamy blue sea of curving, flowing forest." The softness of the landscape emerged only when the hiker stood on the top of a peak after hours of hacking through briars, laurel, and rhododendron. A trip through the forest for one trying to appreciate landscape challenged preconceived notions of beauty. Jeffers noted that to get to the ethereal heights, one had to move up "innumerable ridges, separated by deep and narrow ravines" filled with "impenetrable rhododendron and laurel thickets . . . 10 to 20 feet high." This was the Huggins Hell experience. When James and Richard Huff, sons of a Mars Hill College professor, completed a nine-day hike through the Smokies in the summer of 1935, they attributed the area's "soft appealing beauty" to "thick green vegetation." Yet the thickness was not purely visual; much of their trip account concerned their sense of being trapped by that vegetation. The "iron-hard shrubby growth" punished those who insisted on moving forward.[40]

The fourth step in the path toward flow is to *get better at meeting the challenges outlined in the previous step*—or, as Csikszentmihalyi states it, to "develop the skills necessary to interact with the opportunities available." By reframing the challenge or repetition of the activity as a set of opportunities, one elevates the process, not just the result, as a test of self. For both sets of hikers, this would relate to the "seasoning-in" process that several older members wrote about. As they celebrated their accomplishments in photos, reports, and gatherings, they also celebrated the physical suffering that had gotten them there. If they believed their own copy, that the "terrific fight against elemental obstruction" could be recast as one of the joys of hiking, then they were getting better. Pain memorialized meant that they had sloughed off their city and town skins in the name of their gritty goals. When Fink wrote that he embarked on a crash course in hiking and camping techniques in order to "take some of the toil and trouble" out of the experience, he alluded to this process of heightening the ordeal by getting better at the hard parts. But getting better mostly required time on the trail, as hikers who were "fresh from office work" bent under heavy backpacks and slowly pushed themselves up steep slopes.[41]

Though most hikers ventured into the woods with others, the physical strain of the trail frequently turned the traveler inward, focusing on the pain of the journey or the routine of putting one foot in front of the other. Club members valued the lack of contact they had with others while on their trips. It mattered that they were the only ones out there at any given time. Although they were not Luddites, the intrusion of sounds from automobiles or logging operations could diminish the aesthetics of a trip. They were highly social within their group, but while hiking, they wanted few reminders of the outside world.[42]

The practice of hiking for hours through difficult terrain with the equipment available in the 1920s and 1930s entailed a dozen or more separate tasks that had to be incorporated, sometimes seamlessly, in order for hikers to experience "oneness" with the trail. Travelers in the Smokies had to read maps, find or make trails, scramble on all fours up particularly steep sections, judge their footholds, assess alternate routes, estimate walking pace and the time to their destination, and remember the route they had taken since the hike began. All the while, hikers' recollection of the trail behind or in front of them had to include an index of rest sites, water sites, shelters, hazards, and fords. Sensory studies demonstrate that expert hikers have a greater capacity than novice hikers to remember highly functional landscapes, suggesting that this indexing skill derives from practice: the memory of using certain natural features in the past to avoid trouble or increase comfort influences future decision-making.[43]

With packs averaging 60 pounds apiece, the nine men on the August 1932 end-to-end hike learned on the very first hill that they needed to shed weight. They gave personal items—relative luxuries—to the day hikers who would turn back after a few hours. The doubts that Campbell confessed to feeling 2,500 feet below the first peak were a testament to the seasoning process that had to occur on each hike of more than a few days. Lack of water became a pressing problem by the end of the second day, and when the group was not in the thick tangle of rhododendron, they were scorching in the August sun. They eventually sent the youngest member of the group, the newspaper editor's son Herrick Brown, down off of the ridge to drag water back up.

This was the first moment of danger in Campbell's narration of the trip, and the success of Brown's mission allowed the tale to continue to the point where the hikers had to find the exact spot at which to turn off the main trail. To miss the turn would mean an errant drop into Tennessee and a morale-killing climb to retrace their steps. By taking extreme caution and comparing notes, the group stayed on the correct trail. Day five found the hikers on the 10-mile stretch of the Sawteeth, with continual ups and downs and a narrow ridgeline that dropped away sharply on either side. By this point in the hike, the club members were broken in. The rough walking no longer seemed extraordinary; their bodies were weary but toughened by the strain. Day six was the first "long day's hike," more than 10 miles between camps. That kind of pace allowed the men to think that a thru-hike was, indeed, possible. The Sawteeth were a "delightful bit of rugged wilderness" after five days of trailless, thirsty work.[44]

Difficulty met head-on led the way to an optimized experience. John Muir described a state of flow while hiking as a by-product of the difficulty of steep, slippery slopes. Mountain hikers who sensed the danger of the route ahead saw little of the surrounding beauty but reveled in their own precise movements, "so keenly and narrowly concentrated" that they felt an aroused state of sensation. This was not a transcendental experience wrought by a higher power, but rather a physical experience ignited by strenuous conditions and one's near inability to believe that the goal was within reach. One of Stoddard's consistent themes was the seasoning of the walker. "The first day of a long tramp," he wrote, "is often not the sweetest. Either the shoes do not fit to a nicety or the straps gall and the pack is not well-balanced—and one always wears too many clothes. But more trying even than tender feet or other physical discomforts is the difficulty of mental adjustment." Stoddard explained that day two would not likely be better; that was when the trail tested hikers' determination and made the previous day's aches truly hurt. Day three, Stoddard reckoned, was the end of the "initiation" process. Though a *New York Times* reviewer described Stoddard's take on walking as "a bit too mawkishly moral" for the automobile age, that was perhaps what drew Campbell to the book. It gave the club members from city and town a framework through which to understand their endeavors.[45]

Flow worked as a form of expertise because these professionals from Tennessee and North Carolina became evangelical about their achievements within the clubs, but also through the pages of their published guides and media appearances. Gaining new skills took walking, and more walking took training in how to endure. Every time the main organizers of the hiking clubs signaled to their members what they should think about their own efforts, they made an argument for the leadership's expertise. Every photograph that revealed a sight or area that had once been considered too remote or impenetrable made a case for the specialist knowledge and physical skill that the photographer must possess. The SMHC's schedule of hikes for 1932 alerted members in this way about a trip to the top of Mount LeConte: "You will agree that this is a strenuous hike after completing the 13 miles of the round trip. Only the most capable should attempt the hike down Roaring Fork, as it is quite rough. The leaders know the trail and fully realize the hardships involved." The leaders, Herbert Hunze and William Stanley, were an X-ray equipment salesman and an entomologist. What they knew about hiking they had learned over many weekends.[46]

When the NPS and local hiking groups built trails, they hoped to take some, but not all, of the difficulty out of the passage between trailhead and vista. They focused especially on the difficulty of achieving a rhythm, a feeling of effortlessness despite their hard work. Those who were used to the terrain, particularly those who had been born in and around the mountains, were said to develop an ease of movement that eluded the novice. Regional writer John Judson described local residents moving with a steady, sinuous gait up mountains, passing "outlanders" who clung to trees and rocks. The mountain writer Horace Kephart described the walk of a woodsman as a "rolling motion, his hips swaying an inch or more to the stepping side." Achieving this fluid style was easier with the aid of trails, but the club leaders looked for a sense of effortlessness amid staggering effort. Even on a trail, hikers encountered several hours' worth of briars and rhododendron when moving north from the parking lot at Newfound Gap. And that was fine for some; Broome wrote to the park superintendent that "there are a few mad souls like myself who still take an almost savage delight in fighting those briars for a whole day at a time to achieve three or four hard-won miles." These difficulties thrilled hiking club leaders and allowed them to define their practice as more than a scenic venture. Meeting the challenge of the mountains meant learning to believe that "hanging on by fingernails and eyebrows, clawing for toeholds in the slippery earth," was a normal state of being.[47]

The final step of the flow process, one that could keep the activity going well past the time when it seemed to have been "mastered," was to *keep upping the ante with more elaborate or demanding goals*. The border-to-border crew could push for a longer hike, choose a more strenuous path, or attempt to complete the original route in less time. In the summer of 1933, to celebrate the nerve that it took to complete the previous year's trip, the SMHC leaders opted for a shorter but more technically punishing "vacation hike" that was virtually trailless. "Think of what you can tell 'em when you get back home!" their guidebook gushed. The Huggins Hell crowd could always find newer, darker, more "inaccessible" spots into which they could scrape and plunge. The density and scale of the Smoky Mountains, both on high and down low, meant that even the most active hikers could hardly exhaust the regional possibilities. Photographs, trip reports, and stories shared in annual guides and at meetings gave hiking colleagues incentive to do more, to add their own entries to the roll of accomplishments.[48]

The ethos of achievement never seemed to become heated competition within or between the hiking clubs. Instead, the members deflected questions of superlatives into collegial rivalries like whose boots showed the most wear or whose physical appearance became most unkempt after a trip. The "feral, savage look" that most hikers sported after a few days in the mountains was proof of designed punishment. The SMHC thru-hikers of 1932 played with this marker of achievement. They arrived at Newfound Gap on their sixth day, thirty-six hours ahead of their announced schedule. Here, as Campbell told the story, they faced their next significant challenge: to stop or to press on? Although some members of the group suggested that they take a night in a motel in order to ease their aching bodies, they were outvoted by those who said that to do so would mean the end of the trip. They would be "contaminated by civilization" and never complete the 72 miles in one go. Avoiding accustomed comforts in order to stay rough became a more abstract way of thinking about the goal of extreme hiking.[49]

And so the SMHC group left Newfound Gap on foot, continuing along the AT to camp 2 miles away from the road and away from temptation. The rest of the trip, three days of hiking 38 miles, passed in relative ease. The path was improved, the section of the trail was well known to the hikers, and their bodies were used to the rhythms and rigors. The men lined up and walked as one to the highway that marked the end of the hike. They hitchhiked to a nearby lodge to call a bus company to pick them up. Both in the car and on the buses that took them back to Knoxville, Campbell recalled, their appearance and smell after nine days of hiking repelled the drivers and passengers. But there was joy in such a reception, because their beards, tans, and odor told the tale of hard work in the mountains. That type of appearance became a token, and it heightened expectations of what club members could do to earn it. Stoddard had warned readers that their insistence on walking distances would lead them "to be considered singular," but he counseled that this should be a marker of pride. The club elite took this advice to heart.[50]

Achieving a state of flow was a process, and appreciating that state—even proselytizing it—was a step that upped its aesthetic and kinesthetic value. Each backpacking trip featured an opening day or two of physical suffering that hikers attributed to acclimating their white-collar bodies to the hardships of the trail. And each day in the Smokies typically started with an arduous climb up to a ridge from a low campsite chosen for its proximity to water. So the experience of a long hike was a series of attempts to deal with shaky legs, sore backs, and burning shoulders. Add any side trip off the most-used trails, and hikers could expect to encounter "rolls of man-sized saw briars" and a host of other impediments. If, as Stoddard claimed, veteran hikers knew that the "sheer joy of the journey" was the motivation, then expertise in the mountains was a matter of enjoying the difficulty. Finding rhythm where there was only obstruction and feeling ease amid great effort were signs that one could claim a special body of knowledge.[51]

Modest Opportunities

When the NPS implemented a trail system in 1932 to help visitors choose their hiking experience in the Smoky Mountains, the hiking clubs' enthusiasm was tempered by fear. The NPS created class-A trails (4 feet wide and never more than 15% grades), class-B trails (narrower, steeper, and rougher), and class-C trails (barely visible and extremely rugged). Carlos Campbell noted that many of the class-C trails were falling into disuse by 1937, although some people preferred them to the "beaten path." In the winter of 1934, the SMHC's Guy Frizzell wrote to NPS engineer Robert White about plans for the Mount LeConte ascent via Roaring Fork, one of the roughest paths remaining. Frizzell noted that four trails already led to the top and that club members wanted no more. The high point of the Smokies might become too accessible. SMHC members volunteered to rework the paths, hoping to leave an element of ruggedness in place. Superintendent J. Ross Eakin denied these maintenance requests; NPS planners carefully calibrated their level of manipulation in specific sections in order to design hikers' experiences. The general public must be considered.[52]

Hikers' peak narratives that ended in triumph tended to dramatize the final few feet of the climb, when effort turned into respite and grandeur. Harvey Broome, "weak from hunger and fatigue," reached the top of Mount LeConte in 1920 after scrambling through thickets of undergrowth. "When it seemed I could go no farther," he wrote, "I dragged myself over a low ledge and found we had reached the top." When hiking through the Smokies, many people found that the time spent at overlooks rewarded them for the trip. When telling tales from the trails, on the other hand, they might find that their audiences liked to hear about the trials and errors that made hiking adventurous. It was a decided minority of hikers who enjoyed the toil of hiking through rough landscapes. When the SMHC leaders complained about the improved state of trails in the NPS era, they did so to preserve the grueling experience for future hikers. They had suffered in the name of achievement, but would others have the same opportunity? Would others *want* the same opportunity? The SMHC guide for 1927 encouraged hikers to push on: "The terrific fight against elemental obstruction is one of the joys of this particular trip. No matter how dense the underbrush—keep a lookout ahead." Four years later, the SMHC answered a question that the club's leadership must have heard often: why? Paul Fink admitted finding inspiration in the "atavistic trace of the old pioneer spirit that actuated our forefathers to leave the comparative comforts of their cabins in the lowlands and strike toward the crests of the high ranges they could see boldly silhouetted against the western sky."[53]

The language they used to dramatize their efforts made the hard play of hiking a sort of reenactment of the practice of pioneering. Travel writer George Dacy sold the wonders of Smokies hiking in 1928 as a historical experience: hikers appreciated the "twang of pioneering perils" as they started "roughing it in the undefiled open country under somewhat similar conditions to those that our hardy forefathers knew so well." "Hardy" became a foundational term, a means of conflating youthful exuberance and work ethic. A CMC brochure noted, "There is a reverential regard in most of us for our hardy forbears who hewed their way through the wilderness. In early youth this shows itself in the instinctive desire to emulate the example of these doughty pathfinders. In response to this call of the blood it is the endeavor of the club to offer modest opportunities to keep alive the desire, not only during but particularly beyond the period of youth." These "modest opportunities" were organized attempts to take the ethic of accruing professional experience—an ethic forged in the offices of Asheville and Knoxville—out into the woods. When consecutive days of hiking turned them "grimy and trailworn . . . shoes wrecked beyond repair, clothes half torn off by the clutching fingers of the brush," club members celebrated their appearance to those friends who took them in for a shower or a meal. Those who reached the distant peak, despite the physical nightmare of "uncut mountain fastness" that preceded it, suggested that mimicking the mythical pioneer was not child's play if the life of a pioneer was understood as a lost form of engagement with landscape. Even (or, perhaps, especially) a bank clerk could strive for that.[54]

This ever-escalating sense of what was possible had an expiration date. By the time they were in their forties and fifties, Broome and Fink admitted that they no longer found joy in fighting through rough woods for miles on end. Maintained trails had become more appealing to them by the 1940s, but while they were in their twenties and thirties, club members savored the chance to discover satisfaction in work-like leisure. Fink estimated that an eight-day trip through the Balsam Mountains in the summer of 1924 was the most strenuous ordeal of his entire life, yet it was also "a most enjoyable one." Trapped for days in a stand of blackberry briars, he reported a sense of exhilaration in managing to keep going. The eventual view from Sugarland Mountain was good enough to repay the "wearisome, heartbreaking labor." In his memoir of two decades spent hiking through the southern Appalachians, Broome imagined again and again that he and his companions sought something beyond escape. He wrote, "If the woods were sentient, and some seedling should say, 'Here comes a man,' the whole forest would answer, 'What of it?' " Hikers of the SMHC and CMC cherished the notion that the woods did not care about them, but instead established the harsh limits within which they tested themselves. The physical struggle allowed hikers to boast of their tactile sense of landscape, their knowledge-in-action. They had learned things that others could not, at least not without great effort.[55]

On one of the last trips he led for the CMC, George Masa tried to steer his hiking party away from a place of great difficulty. He wrote in his trip report from May 1932 that he had suggested dropping off the Smokies' state-line ridge at a point that would have connected the group with an NPS trail to take them back to camp in a roundabout but easier way. They had already fought "briers and snags" enough that day, and he doubted the resolve of some of the thirteen other hikers. But the group overruled him, so they plunged down into what he told them was "called HELL." His advice for future leaders making the trip was to follow the more moderate NPS path, but he admitted that "it was a good experience thru one of these 'HELL': Rhododendron, laurel thickets with greenbriers." With those words for his fellow hikers, Masa captured the pattern by which lowland professionals repeatedly put themselves into the stern grip of circumstance and convinced themselves that they liked it.[56]

A Priceless Asset

Marion Pearsall in Isolation

"Back of Beyond"—that is what the writer and outdoorsman Horace Kephart admitted he sought when he moved to the Appalachian Mountains in 1904. His *Our Southern Highlanders* presented the mountains as a magical place where the world-weary could find something new. Much the same could be said of McFarland, MacKaye, and the hiking community of the Smokies. They longed for a "Back of Beyond" to which they could escape on weekends. The remoteness of their chosen mountain destinations allowed them to think of their corners of the Appalachians as essentially unpopulated; people mattered less to them than the views and the conditions. Kephart diverged from those figures with a second motivating factor that drew him to Tennessee: he noted in 1922 that he had wanted to be around "a people that had the charm of originality." The mountains offered certain attractions to travelers looking to get away from people, but they also provided new ways to connect. The "charm of originality" promised on high was an antidote to the tedium of accustomed society. The newness of the mountains could be pleasant, frustrating, or downright antagonistic, but Kephart claimed to embrace it all.[1]

University of Kentucky medical anthropologist Marion Pearsall would have empathized with this quest for authenticity in others, even as she might deny that her travels put her in touch with anything unique. She wanted contact and engagement with mountain people as a way to examine how social groups interact and understand change. Her work in the late 1950s and early 1960s followed two paths in the mountains, because she hoped to teach colleagues about Appalachia in two distinct ways. First, Pearsall contributed to the social scientific literature on insulated mountain populations by studying how residents combined homegrown medical knowledge with imported health practices. The insight that she conveyed to colleagues most frequently was that the people of Leslie County, Kentucky, were highly adapted to the material constraints of their surroundings. From the flexibility of their "general tendency to live from day to day" to the specific ways in which they described their neighbors, cared for the ill, or traveled from here to there, her subjects displayed habits that worked for them in local contexts. All were shaped by local topography and infrastructure. This was not the environmental determinism popular among geographers in the 1920s; she did not think in terms of "bad" land creating "bad" people. Rather, she encouraged colleagues to think of the physical setting of Appalachian Kentucky as a dynamic force that mixed with economic arrangements, kin and neighbor networks, and political commitments to define the limits within which residents lived. Locals' experiences might be determined to a great extent by these slowly shifting structures, but they could also act to change the structures more rapidly.[2]

Second, Pearsall thought reflexively, making the presence of the social scientist in the mountains a crucial part of her analysis. She rejected any notion of stasis in the hollows of Appalachia—she did not believe that things were now as they always had been in eastern Kentucky, nor did she believe that missionaries, mining and lumber companies, and researchers like her were able to sequester themselves from the people with whom they worked. Pearsall argued that multiple scales of social and cultural change happened simultaneously in Leslie County and that professionals who were interested in directing change in certain directions had to pay attention to their own role in the process. There was a day-to-day quality to her sense of how local worlds shifted. She encouraged the great sensitivity needed to examine the swirling, competing changes happening within a place that seemed utterly static to nonresidents. Training oneself to notice what laypeople could not was a thorny proposition; it required a specialist's sense of movement in multiple dimensions. Not the least of these dimensions was the collision and collaboration between residents and nonresidents. Her second contribution in the early 1960s was providing a schema for measuring and channeling the flux. The technique that she used to figure out the contours of the place was generalizable, even if Leslie County was not. Pearsall observed that "patience, education, and mutual trust" were the pillars of progress there.[3]

One of the most pronounced tensions in the work of researchers studying Appalachian communities in the fifties and sixties was the need to understand the region as a sum of its parts versus the obligation that some felt to treat each locality as its own complex of culture, environment, and economy.

The first mode of analysis—the *"if you've seen one hollow, you've seen them all"* approach—was adopted by people bent on improving the lives of mountain residents. They saw nothing wrong with essentializing Appalachia, if doing so allowed policy makers to grasp the extent of the challenges posed for the region's population. The second mode—the *"every hollow is unique"* approach—resulted from ethnographic attempts to contextualize communities in space and time for other outsiders, whose careers required them to gain access. Taking cultural systems seriously meant refusing to simplify them for the sake of brevity. The field notes of a trained observer could fill volumes with a glut of indispensable detail.

The two frameworks challenged Pearsall. If any mountain community, properly analyzed, could stand in for the rest, then far-reaching change could be instituted through efficient research. But arguing that any given community stood in for the rest undercut the privileged position of the participant observer. Social scientists like Pearsall depended on the professional credentials that they earned by translating inscrutable mountain locales for colleagues. Her published work and conference presentations were, after all, intended for anthropologists and sociologists. Fieldwork gained the anthropologist professional capital with an emphasis on the hurdles that she overcame. Yet focusing exclusively on the unique aspects of a given community threatened to make local findings useless for planning and policy making. Pearsall was devoted to action research, a methodology that emphasized the role of expert consultants advocating change. She wanted fellow change agents to apply her findings, but she also wanted to demonstrate that she had achieved something exclusive. So in Pearsall's analysis, her target sites became shape-shifters, changing from generalized places that yielded useful insights for social scientists to "unique and personal" places that only revealed their "flavor of life" after extended digging. Her take on mountain lives circled the very question of what it meant to consider regions and the skills necessary to carve out an effective role in them.[4]

Snooping Around

Pearsall was born in New York in 1923. The daughter of a mechanical engineer and a schoolteacher, she lived most of her childhood in the semirural heart of upstate Madison County. Her academic training took her from undergraduate studies at the University of New Mexico, where she double-majored in biology and anthropology, to the University of California at Berkeley, where she received a doctorate in anthropology. The faculty at Berkeley during

Pearsall's graduate work was filled with leading figures in the ethnographic and curatorial worlds, including Alfred Kroeber, R. H. Lowie, Theodore McCown, and E. W. Gifford. The ethnographic fieldwork that produced her dissertation took her in 1949 to Black Gum Gap, Tennessee, where she focused on the experience of change in small communities. Eastern Tennessee provided the half year of participant observation that she needed to generate specific knowledge of how people comprehended change. Between mid-July 1949 and mid-January 1950, Pearsall lived on the eastern fringe of Sevier County and interacted with roughly 120 local residents of the area's mountain creeks and remote ridges. She revised the dissertation while teaching at the University of Arkansas and the University of Alabama between 1950 and 1956. Next came a two-year stint as a postdoctoral resident with the Russell Sage Foundation in Boston, where Pearsall worked in hospitals to study social organization among nurses. When the nursing research ended, Pearsall was ready to put her ideas about social life to the test in a new research site. Eastern Kentucky called.[5]

After *Little Smoky Ridge*, based on her Tennessee work, was published in 1959, Pearsall shifted her focus to research in a mountain community 80 miles north of Black Gum Gap. She began a "social anthropological study of health and culture" in Leslie County. In April 1959 she informed Rufus Fugate, the University of Kentucky's agricultural extension agent for the county, that she would be "snooping around" that summer. The 1949 fieldwork was on her mind as she set her sights on Leslie County. Several themes from Black Gum Gap carried over to the new site. First, she believed that the social conditions of both places were an effect of "local topography, climate, and traditional ways of adjusting to the natural environment." Context mattered, and the context that mattered most to Pearsall was the material world's power to influence—but not determine—human behavior. Second, in both Sevier and Leslie Counties, mountain residents experienced as processes that which mainstream Americans understood as things. This was particularly true of time and nature, two concepts that Pearsall explored as key elements of social change. To the people with whom she interacted, time and nature were fluid, social affairs; it would make no sense to separate them from the social worlds in which they were experienced. Third, both mountain populations experienced "cultural abrasion" that impaired individuals' ability to find stability in life. This cultural abrasion could result from contact with nonresidents or through internal dynamics that pitted residents against each other. She entered Leslie County with high ambitions, attempting to understand nothing less than what a local medical worker later called "bio-psycho-social-economic problems."[6]

Perhaps most of all, Pearsall took with her a more measured view of what could be accomplished through social science. Her time in Appalachian Tennessee had shown her that becoming an insider was an illusory goal, despite the scholarly literature that spoke of field-workers "going native" and "passing the point of field rapport." The people who lived around Black Gum Gap did not form a homogeneous culture that functioned as a buffer against the outside world. Instead, local residents divided themselves by topography and geography. Residents of the relative low ground along Webb Creek, who were closest to the roads that led to towns, viewed the people who lived just to the north on the ridges as "backward and primitive." The creek dwellers spoke of "lazy" people "back on the mountain" and wondered if there was anything that could be done to pull them out of their torpor. Pearsall was careful to use the term "neighborhood" when referring to Black Gum Gap; she distinguished this collection of people living in general proximity to each other from the more formal and loaded term "community." Pearsall noticed pronounced "lines of antagonism" running between neighbors in Black Gum Gap, where family unity trumped all and became imprinted on the rough landscape. Short-lived cohesion among the people of "Little Smoky Ridge" emerged only when a "stranger enter[ed] the foot of any branch." Children typically relayed the news of the visitors' presence up the hollow, and soon everyone had prepared themselves for a knock at a door or a honk of a car horn.[7]

Shifting to the north ten years later, Pearsall noted that Leslie County was not a unique collection of mountain terrains and communities, but rather an amplification of attributes found in ready supply elsewhere. It was "more rural, more isolated . . . and more impoverished" than other counties in Southern Appalachia. It had a more pronounced set of access issues that she found to be at the heart of public health challenges in the mountains. For residents of Leslie County, the forbidding topography, the crazy quilt of roads and trails, and the "tortuous distances" from homes to institutional services determined their sense of what was possible. Pearsall never thought that she might become one with them.[8]

Before she embarked on the new project, Pearsall contacted an established Leslie County expert, Mary Breckinridge of the Frontier Nursing Service (FNS). Breckinridge had founded the FNS three decades earlier to provide health care to those without any access to trained medical professionals. By the late 1950s, she controlled a quasi-independent, charity-reliant public health system with four nursing centers in the county. Breckinridge wrote back, saying that she looked forward to Pearsall's visit and thought that it would "save us both time" if she read Pearsall's book and Pearsall read at least chapters 13 and 17 of her book. Breckinridge had completed *Wide Neighborhoods* seven years earlier, and it served as both autobiography and institutional history of the FNS. The recommended sections told of her mountain horseback rides in the summer of 1923 and her thought process two years later, when she selected the county as the site of her life's work. The chapters also made a case for Breckinridge's mastery over the area's adversities. Hard travel, bad beds, and an almost total absence of "calendars and diaries and marriage lines and birth certificates and even clocks" challenged newcomers. Breckinridge seemed to be saying that if Pearsall could imagine the trials that the FNS founder had faced over the years, she might get a sense of what it took to affect real change there.[9]

That was not all that Pearsall studied; she prepared a "background reading" list of forty-four books and articles and got down to work in the spring of 1959. Several sources seem to have been crucial to her thinking about Leslie County. Chief among them was Robert Redfield's 1955 book *The Little Community*. Redfield described the "isolated mountaineer peoples of the Appalachians" as exclusively family focused and lacking "organized group life." Instead of a community, Redfield argued, each hollow in the mountains was a "world to itself." Researchers would find it difficult to access those worlds, for even neighbors did not think of each other with an expected sense of intimacy or attachment. Pearsall typed forty pages of notes on *The Little Community*, focusing much on chapter 6, in which the anthropologist tackled the issue of insiders and outsiders. He identified the "central problem" in ethnographic research as the need to manage insider and outsider viewpoints in relation to each other. Scholars first had to adopt an insider's view in order to understand states of mind, social structures, and the like. But making sense of that for other scholars was worthwhile only if they could relate the insider views to the intellectual preoccupations of colleagues. The process was further complicated by ever-present change. Even insider viewpoints shifted constantly, Redfield argued—after all, they were just a "temporary construction, a precipitation of a crystal from thoughts that from day to day are carried in the flowing solution of life's doings."[10]

The other main source of inspiration came from James Brown's work in eastern Kentucky in the early 1940s, synthesized in his Harvard dissertation but not published until 1988. Brown's dissertation would have confirmed for Pearsall that getting a foot in the door of Appalachian Kentucky took strategizing. Brown had chosen his Beech Creek site in Clay County in 1941 specifically because he had a "sponsor" in the area. He had to abandon his original choice, a hollow up Cutshin Creek in Leslie County, when the wife of his intended sponsor became ill. Brown's work exhibited the pattern that Pearsall adopted: plenty of detailed analysis of life as lived in the field, interspersed with regular insight on how the researcher could get the most out of his or her time. Pearsall later discussed the role of the social science researcher in the mountains with public health colleagues. The consensus was that participant observers had to be "guarded" in their approach, forming friendships with locals through the "legitimizing influence" of elders or school officials. Those friendships then became the base from which fruitful data gathering could be had. It had worked for Brown almost two decades earlier, in a spot 25 miles from that which Pearsall had chosen to investigate.[11]

She took this advice to heart. When she visited Leslie County families during her stay, she used the nurses from the Cutshin Clinic as her means of making connections. She informed the clinic staff that two visits would be necessary to get the information she wanted: an initial meeting in which she was introduced by a local figure who was trusted by the interviewees, and a follow-up in which Pearsall asked questions and took notes. Breckinridge's approval opened doors in Leslie County for Pearsall, but she also leaned heavily on Fugate for connections and logistics. Johnny Asher, the owner of the local drive-in theater, offered to act as a liaison between Pearsall and residents of the more remote sections of the county.[12]

Accompanied by a research assistant, Pearsall gave an address at a June 1959 meeting of the Hyden Lions Club as an official announcement of her plans. Club member Roy Sizemore, a former postmaster and frequent county official, introduced them as guests of the club, and Pearsall admitted her concern that local people not see them as "more of those reporters." She explained her method as a new approach to community health; public health was the measure of a community's "potential energy for everything else worthwhile in life—for work, for play, for helping others, for worship." She used a little self-deprecation to win over the audience; while the people knew that health was a complex thing affected by much more than processes within the body, it took "us professionals" a long time to achieve this insight, she admitted. Traditionally, federal and state funding had gone to research on soil, crops, and livestock within the University of Kentucky's Agricultural

Experiment Station. That was now changing somewhat, as officials came to see that people were the state's "most important and most valuable resource." She was not an expert, Pearsall declared, but she would be able to study local problems until she could make recommendations for action to the Extension Branch. "My own feeling," she concluded, "is that rural counties have something to teach the rest of the country when it comes to handling health problems. . . . Somewhere there should be a proper balance of physical, social, and spiritual, and that is what we hope to find through studies like ours."[13]

Pearsall's strategy was clear: she presented herself as a nonexpert who needed to learn from locals and as a professional of a field that was well intentioned but prone to overlook the obvious. She flattered audience members by suggesting that they came closer than those outside the mountains to finding balance. *Help me see what you see*, she asked. By presenting herself as a student in the classroom of the community, she followed a standard technique of ingratiation to overcome reticence or antagonism. (When the University of Kentucky's Mary Ann Stillman Quarles conducted her master's research in Leslie County a decade before, she presented herself as a "school girl" and explained to locals that she was "writing a big paper which had to be finished before completing school.") In her fieldwork in Talladega, Alabama, in the early 1950s, Pearsall and her University of Alabama colleague Solomon Kimball had worried about local residents' distrust of outsiders. Acceptance of the researchers was a delicate matter, and it eventually came in Talladega through the "casual, but arranged, situations" in which residents could take the measure of the academics. Pearsall and Kimball described this in 1955 as the "induction of outsiders into the system of informal relationships which gave stability to the 'inner circle' of the structure." In her Lions Club talk, she hoped to kick-start that induction process. She wanted to be visible in Leslie County, but she did not want to stand out.[14]

In Breckinridge, Pearsall had unknowingly consulted a master of the art of staying hidden. Breckenridge claimed that she chose Leslie County in the 1920s because of the near impossibility of reaching and navigating it. "Our inaccessibility," she wrote in 1952, "was a priceless asset." Breckinridge explained that she set herself as an example to other medical professionals: if she could serve a population as remote as the people of Leslie County, then none of her peers could argue that their own plans were too ambitious. Like Breckinridge, Pearsall claimed possession of Leslie County's inaccessibility. It was a way to display the effort required to work in mountains.[15]

Layers of insiderdom formed around the contact between residents and nonresidents, revealing locals' presentation strategies. Mountain people could take advantage of nonresidents' preconceptions, Pearsall found when she became a confidant of those with whom she worked. A county official told her that Breckinridge wanted to keep the central part of Leslie County isolated. The theory was that Breckinridge resisted road construction along the creeks south of the FNS complex in Wendover because she wanted to keep outsiders at bay and because she received donations from people who saw the mountain isolation as a suitable cause for charity. She was afraid, in other words, that connecting Wendover too well would hurt her bottom line. An FNS staff member denied this story, explaining that the site was precarious; rockslides threatened to destroy everything if construction crews came in with heavy machinery. The impulse to keep eastern Kentucky usefully isolated relied on the assumption that mountaineers suffered in their condition and a recognition that sympathy could drive people to spend money.[16]

Whether the Breckinridge rumor was true or not, it is telling that someone on the "inside" tried to bring Pearsall into the fold by exposing how the one-time outsider used outsiders' stereotypes to tend to the medical needs of residents. Pearsall, meanwhile, was typing pages of notes each day to translate for herself and colleagues the everyday details that built insiders' experiences. She subtly presented herself as something closer to an insider than an outsider, at least in terms of her ability to push through the barriers that kept the two apart. She confided in local nurses that she understood how hard it could be to work under someone as powerful as Breckinridge. Saints should never be confused with angels, she told the FNS staff when she returned to Leslie County for a workshop in 1972.[17]

An Ecology of Contact

Many mountain residents might have exhibited a general suspicion of strangers, ranging from a reticence in front of reporters, who would become much more prevalent a year or two after Pearsall visited, to outright intimidation of "furriners" or "outlanders" seen to be trespassing or meddling. This was surely exaggerated by some accounts and understated in others. But Appalachian residents' relationship with the people who visited them changed in the decades following World War II. The effects of unemployment broke down any tight grip on the area which local residents might have once had. By the late 1950s, a slight majority of surveyed mountain residents said they hoped that their children would migrate. It was the promise of work that drove such trends, not a newfound love of lowlanders. Still, the desperation for regular income softened the edges of antagonism that had marked the era when subsistence was possible in the mountains. Participation in the industrial workforce beyond coal chipped away at the distinctions between the residents of the highlands and those of the lowlands. Class antagonism might have had more influence here than regional affiliation. West Virginia researchers admitted that their work with low-income residents in coal communities was hampered by the residents' distrust of people whom they associated with a hypocritical middle class. This tension turned any attempt to establish "rapport" into a struggle to convince locals that answering questions or working with strangers would not open them up to criticism or calamity.[18]

Pearsall's grasp for expertise stemmed from her dedication to action research, a type of applied social science that gained influence among American anthropologists and sociologists in the 1950s. Action research practitioners sought to understand social change in order to generate specific forms of it in particular contexts. Whereas participant observation offered insights into the life of a discrete group, action research layered onto the participant observation method a more pressing role for the researcher. The original model developed by Kurt Lewin, the researcher who popularized action research in American academic circles, presumed a three-stage process of social change. A change agent (1) intervened in a stable social group to produce (2) an episode of action, which ended when (3) the group returned to a stable (yet potentially new) state. Pearsall's mountain work sought what Appalachian Studies scholars have called "viable forms of regional politics." These are ways of making positive, human-focused change in mountain communities while recognizing that the material realities of life cannot be explained as the by-products of myths foisted on Appalachia from the outside. The available social science literature in the late 1950s painted the people of the mountains as a maladjusted, unambitious lot whose problems stemmed from their disconnection from mainstream American life and their environmental shackles. Pearsall disagreed, and she saw action research as a way to nudge residents and policy makers in certain directions, while figuring out what both groups wanted and feared.[19]

In the context of Appalachian Kentucky, pursuing action research meant that Pearsall problem-solved in overlapping ways. For her, a study of a certain topic was simultaneously an examination of the procedures and practices through which the topic was studied. The problem, she believed, was that not enough social scientists were willing to write reflexively about the "shadow" study. In the social worlds of fifties and sixties Kentucky, a researcher like Pearsall took her occupational network with her, in the sense that the expectations, encouragements, and directions of her peers influenced her practice. Any moments of alienation which she felt in Leslie County were layered with her membership in professional circuits designed to theorize that distance. The methodological literature of the 1950s promised her that time in the field would, indeed, come to an end, and that she would be "out and in the company of understanding colleagues" eventually. Pearsall thought that the past two decades of pioneering community studies had failed to capture the dynamic human relationships that made everyday life rich with meaning. Sociologists had asked their subjects structural questions and focused on "things and categories," which resulted in taxonomies of class, age, ethnicity, and institutional affiliation. Anthropologists who practiced ethnographic techniques tried to adapt relatively simple schemas about clans and kin networks to grasp the more complex groupings that they encountered in their fieldwork. What Pearsall encouraged, along with Kimball, was for scholars to observe personal interactions between subjects as they participated in a given event or action. *Look at people in motion*, she stressed, *if you want to see how they actually live.*[20]

Kimball and Pearsall contrasted their work with the community-study sociologists who came before them by using the analogous contrast between ecologists and taxonomic biologists. Like ecologists, they were more concerned with the "network of relations and equilibrium of living forms" than with the characteristics of those forms. Pearsall meant to remedy the stifling focus on structure by using Appalachian Kentucky as a site to teach colleagues about the techniques needed to discover things in the mountains. Some of her Leslie County field notes served as meditations on what residents might have done when nonresidents were not watching. A significant portion of them acknowledged that there *were* nonresidents watching and that more would visit in the near future. Her work evolved between June 1959 and late 1960 to focus on the "behavioral barriers between practitioner and client," wherein the practitioner might be a nurse, a doctor, a county health official, or an anthropologist bent on improving rural health care coverage. Rendering technical meant turning locals' everyday practices within their accustomed landscapes into abstract models of circulation and change.[21]

Pearsall targeted her problem-solving at the health care access deficit in Leslie County and the interpretive deficit in Appalachian social research. The former was due in large measure to what David Whisnant described as the "coal-dominated social, political, and economic system" of Appalachia. Working through the limitations of funding and infrastructure would require residents, government officials, and external corporations to collaborate with change agents like Pearsall. There was a "distinctly patchwork quality" to local health care, she observed, and money and roads were needed to mend the seams. The 1960 Census revealed that the per capita income in Appalachian Kentucky, $841, was 44 percent of the national average and that the local unemployment rate was 76 percent higher than that of the nation. In Leslie County and the other counties lining the easternmost reaches of the Kentucky River, mining operations outnumbered manufacturing establishments by a factor of seventeen. In 1972, the Appalachian Regional Commission counted fifteen hundred mines in the twenty-one counties that made up eastern Kentucky.[22]

The Leslie County that Pearsall canvassed in 1959 was not a caricature of isolation and backwardness. She understood quite early in her fieldwork that it was an area in which movement and connection were difficult but widespread. This was not the Appalachia that Harry Caudill wrote about as a product of a "century of isolation." It was a world inflected by "local" ways of doing things, but one that was plugged into wider economic and political networks. These were the decades when subsistence farming disappeared from the mountains, when young people moved away for entry-level manual work in regional cities, and when the boom-and-bust cycles of mining and logging continued apace. Problems resisted solutions. The previous autumn, state officials had toured the mountain counties to encourage residents to make their garden patches and hillside farms more productive. With words that echoed decades' worth of Appalachian observers' thoughts, Chad Drake deadpanned in *Mountain Life and Work*, "The Undersecretary of Agriculture did not give any concrete suggestions as to how this might be accomplished."[23]

Health practices in Leslie County were an ever-shifting combination of homegrown remedies and professional, medicalized approaches. Residents used "doctor" as an intransitive verb; it was something that they did to themselves when they sought formal aid from a physician. Doctoring was not their first course of action when faced with sickness or injury, and the local medical establishment knew it. Instead, residents "worried" over symptoms for weeks and months, praying for relief from their conditions before seriously considering doctoring. Only when symptoms became incapacitating did formal medical care become a reality for many residents. Even then, they wanted nothing to do with hospitals. Pearsall heard a phrase that she thought summed up the impasse: an interviewee remarked, "They're threatening him with the hospital again." Setting appointments and then getting patients to them was a complex, labor-intensive task. The county health department worked with the FNS and area hospitals to provide a net of medical care to those in dire need. For example, Pearsall followed the case of a disabled child who lived on Camp Creek in the central part of the county. Each time he needed to be seen by a specialist, a multilevel process kicked into gear. The child's family caught a bus or hitched a ride in a car for the 6-mile journey to Hyden, where they were met by an FNS courier. The courier then drove the child and his mother 30 miles to the Oneida Mountain Hospital in Manchester. Doctors in Manchester then made referrals to specialists at the Shriners Hospital in Lexington (130 miles away) or the Children's Hospital in Louisville (200 miles away). Getting the child and mother to those institutions required further logistical work and expense; the Crippled Children's Association paid some of the bus expenses after the first trip. The hospital in Hyden worked solely on a walk-in basis, because appointments were so hard to keep. "Schedules are so inconvenient," noted Pearsall, "and people will have to come and wait around anyway." She paid attention to these arrangements, because they illustrated the tricky routines adopted by people in Leslie County to deal with poverty and landscape.[24]

Pearsall rejected the notion that mountain residents who avoided this kind of care in favor of home remedies had "given up" or were backward. Rather, she pointed to the sheer exhaustion of individual and institutional energy which came from such arrangements. Residents who "worried" over illnesses were not passive in their own eyes; they were following a course of action that usually solved the problem. It was a wonder that anybody received specialized care at all. Six years after her fieldwork in Hyden, Pearsall attended a conference on Appalachian mental health sponsored by the Southern Regional Education Board. One session focused on the so-called Eastern Kentucky Syndrome (EKS), which psychologists believed was caused by the physical demands of coal miners' lives. The symptoms of EKS were general weakness, lower back pain, insomnia, and feelings of being overwhelmed. Conference participants debated the usefulness of the concept, but they seemed to agree that mountain residents identified with it and saw the status of having EKS as a legitimate social role (the "sick" role). One attendee summarized the behavior of those struck with EKS as "various forms of acting out." This echoed the observations of West Virginia social workers who lamented that kids raised in the mountains tended to quit in the middle of group games if their team was losing. An extension agent warned that this type of flight from uncomfortable situations could produce adults who walked off a job "in order to go hunting or fishing." Another observed that nothing definitive should be said about EKS until sociologists, anthropologists, and psychologists had completed "careful follow-ups" of those hospitalized with symptoms.[25]

These warnings both fit what Pearsall learned in Leslie County and missed the mark. On the one hand, she observed residents' general resignation to poor health. Local people accepted "a chronic state of 'dullishness' as natural" and took steps to correct it only when they seemed to be on the brink of death. Those emergency moments cost time and effort that could have been spared through prevention. But no one in Leslie County seemed interested in long-term planning. The extremes of the place advised against attempting to manage the future. On the other hand, Pearsall marveled at the "eclectic" nature of local medical practices and viewed the impasses that occurred between medical professionals and residents as symptoms of typical cross-cultural misunderstanding. Doctors and nurses did not possess an "automatic halo" in the mountains, she observed, but locals' wariness resulted more from their defense against "modern ridicule" than from ignorance or defeat.[26]

Pearsall planned a book titled *Healthways in a Southern Appalachian County* which would replicate the intimacy of *Little Smoky Ridge* and apply it to an analysis of the overlapping systems of medicine and health. The book would touch upon the difficulties of gathering data in a place like Leslie County, especially the "problems of participant observation in this culture." She stressed that the mountain terrain and the settlement pattern of "open country neighborhoods along numerous watercourses and narrow valleys" were the bedrock factors that produced local healthways. The logistical obstacles that thwarted transportation and communication did not result in surrender, as many analysts reported, but rather a highly adapted set of practices created within "small, atomistic neighborhoods." Increased contact with external markets resulted in the "distinctive almost 'folk' culture" breaking down into a less distinctive "'lower class' culture with some 'middle class'" elements. The breakdown was to be expected, because it was a by-product of the interventions that outsiders had attempted in the mountains for decades.[27]

When her first round of fieldwork in Leslie County had ended, Pearsall presented a paper at the annual meeting of the American Anthropological

Association in Mexico City. It served to summarize her thinking at that time about the role of academic consultants in "action agencies." She reiterated the point that she had made in print five years earlier: social scientists were better at analyzing social structures than they were at making sense of fluid interactions between people in the midst of "contact" situations. Pearsall took the point further, introducing "change agents" as the main players in these moments of contact, executing "manipulative actions and interventions." None of this was sinister or negative; it was an open-eyed assessment of how change agents caused ripples.[28]

Pearsall wanted to study not just change in a place like Leslie County, but the "changing" that was attempted by entering health officials and scholars. The model that she developed by the time of her American Anthropological Association talk focused on the shifting nature of five variables: the cultural, social, psychological, physiological, and physical states of the two systems engaging with each other. Pearsall acknowledged that the psychological and physiological aspects were beyond the expertise of the average anthropologist; work along those lines would require interdisciplinary collaboration and much more money than she had available for her project. But for the cultural, social, and physical aspects of the systems, her model promised to focus attention on the give-and-take that occurred during the "more or less conscious manipulation of situations in action programs." Charting "contact and change" between the systems meant mapping the *physical frontier*, an "ecology of contact, an arrangement of persons and things in physical space"; the *social frontier*, a "sociology of contact which may be conceptualized as an inter-cultural network of statuses and roles stemming partly from the separate traditions and partly from the contact situation itself"; and the *cultural frontier*, where "the accumulated bodies of knowledge, customary techniques and procedures, and traditional value-orientations of the parties meet, and to the extent that these differ, they pose dilemmas of choice for all concerned."[29]

Pearsall's model rendered Leslie County as a technical space in which sensitive, nimble specialists operated. It was an abstract way to conceive of one's procedures when studying residents, their environment, and their interactions with nonresidents. It was also a testament to the work of an expert like Pearsall. It had taken time and acute perception to see multiple scales of movement occurring at once. She put this model to work in an article published a year later in the *American Review of Respiratory Disease*. Tuberculosis was the subject, but again, the topic served mainly to shed light on the processes through which public health officials and researchers attempted to change behaviors in the mountains. She was in the business of studying systems, but lest any of her readers think that "system" implied stasis, Pearsall explained that she was concerned with a "system of interrelated parts changing through time in response to internal and external forces."[30]

Major differences in viewpoint between residents and change agents made it unlikely that locals would accept the guidance of visitors. Positive change might still occur, Pearsall suggested, if the process through which agents presented their advice fostered supportive social relations. The regularity and style of meetings, the number and manner of the agents, the style of the presentations, and cross-cultural dialogue about the presentations could salvage action programs that seemed doomed to failure. This was Pearsall's metacommentary on the experience of fieldwork, a report on the practice of intervening in Leslie County.[31]

On the Move

The cusp of the 1960s was a time of rapid, uneven change in and around Hyden. It was a time when the houses on one side of Camp Creek had electricity while the homes on the other side, dozens of feet away, had none. It was a time when the number of individual farms was cut in half but the number of working trucks on the county's farms tripled. It was a time when the number of telephones increased over five years from six to eighty-one. Margaret Motter, a settlement worker from neighboring Harlan County, had observed a decade earlier that eastern Kentucky was a mixture of nineteenth-century ways and contemporary developments. Motter nodded approvingly at the roads that had been built to accommodate lumbering operations and the Blue Diamond mine complex. She was encouraged by the presence of a radio in one house in Big Rock and a woman's desire to have her kids go to school. In the same vicinity, however, Motter saw windowless cabins of "crude construction" and an old woman smoking a pipe. Bus service had started in some areas of Leslie and Perry Counties, as had the spread of telephones and electricity. Much of what Motter called improvement came via roads, whether paved highways connecting local businesses to the national rail system or sinuous dirt roads that county workers sporadically patched and graded.[32]

Pearsall, too, saw roads as the key to Leslie County's modernization. And though she never wrote about it directly, mobility on mountain roads became the primary way she experienced Leslie County. It should come as no surprise that she concluded that residents were part of a dynamic system of interrelated fields of action, given that her work exhibited the constant motion of daily life. She got around, and everyone she spoke to in Leslie County got around, too. Equating poverty with physical inertia was simply inaccurate. Pearsall wrote, "Whatever level-of-living scales are applied (income; all-weather road mileage; percentage of homes on all-weather roads; percentage of families with automobiles or trucks; condition of housing; percentage of homes with electricity, bathroom, piped water, telephone, radio, television, and the like), Leslie ranks in the lowest 10 percent of Kentucky's one hundred and twenty counties." But those figures could change, and with them could change the shape of local lives. "It's interesting," Pearsall wrote one month into her fieldwork, "how much of the modern world is hauled back into the mountains by truck, jeep, and even by sled today." This process had started decades before. University of Kentucky graduate student Mary Ann Stillman Quarles had worked as an FNS nurse for two years in the late 1940s, and the experience convinced her that local residents were in the midst of an unprecedented integration into national economic and social networks. For Quarles, the factors that mattered most were roads, distant jobs, the increasing availability of credit for consumer purchases, and the thrust of outside organizations with reform agendas. Most of the connective activity that Quarles identified was associated with the growth of large-scale coal mining in the region after World War II.[33]

Pearsall's stay in 1959 coincided with a boom in regional coal mining, *the* economic life of Leslie County. Although many mines shut down for long stretches in the fifties, that summer local companies received new orders for coal which kept the mines running for the better part of a year. With the full-time employment came increased take-home pay, which in Leslie County meant increased mobility. The percentage of local residents who owned functioning cars jumped from 19 percent to 41 percent between 1959 and 1964. The number of trucks doubled, too. Change was characteristic of the coal camps of eastern Kentucky. Population shifts were constant, and the roads leading from the mines connected the residents to the lowlands well before other mountain communities experienced such links. When James Brown had looked for a suitable place to launch an investigation of a mountain social system in the early 1940s, he avoided the coal camps as a result of this flux.[34]

Not so for Pearsall, who actively sought a place in which change coursed through residents' lives. Their roads and their mobility became tangible vehicles for her study, and early in her stay she found that physical access to institutional health care was a major barrier. Pearsall developed several ways to think spatially about Leslie County. In addition to her extended "frontier" metaphor, residents' worlds could be viewed as a series of concentric circles delimiting the distance they traveled for care. They were enveloped in 1-mile circles (the average distance to a mission nurse), 6-mile circles (the average distance to a mining company doctor), a 17-mile circle (the distance to a doctor and nurses in Hyden), and a 27-mile circle (the distance to the "more complete service" of a Hazard hospital). By the end of her first month in Hyden, Pearsall wrote as much about travel times and routes as about medicine. This was not a complete departure for her; the last paragraph of her dissertation had issued a wistful observation about Black Gum Gap: "It is a telling commentary on social programs that one road may be expected to accomplish what thirty years of missionary work has not." The achievement was to encourage locals to act more like nonlocals, to think about their jobs, houses, and futures with more ambition. Mobility seemed to make people think of new destinations.[35]

Unlike the Smoky Mountain hikers, who organized their mountain experiences via peaks and ridges, the residents of eastern Kentucky thought in terms of rivers, forks, and branches. Pearsall's research zone was a portion of the Cumberland Plateau framed by the 260-mile-long Kentucky River, with its three finger-like forks joining 60 miles northwest of her rented room in Hyden. The river's headwaters sat 30 miles to the southeast, along the slopes of the 125-mile ridge called Pine Mountain, which had traditionally served as an obstacle to east–west travel. The hundreds of creeks that splashed across Leslie and Perry Counties fed into the Kentucky River's North, Middle, and South Forks. Hyden sat on the Middle Fork, and that waterway, along with tributaries like Greasy Creek and Cutshin Creek, governed Pearsall's itinerary. "Commodity Days," when county officials distributed staples to the 1,524 families that qualified for public assistance, were organized by creeks. Waterways served as basic, if tangled, orientation devices; the owner of the café in Hyden identified the man who made her counter stools as the man who lived "way up one of the rivers here." Roads followed the largest waterways, and human habitation followed the smaller creeks and branches that splintered up and away from the roads. This "dissected plateau" zone suffered from what the Presbyterian minister Jack Weller called "compartmentation" or the "environmental problem" of Appalachian valleys in 1965. Folklorist Leonard Roberts described this area in 1959 as "the most isolated acres of the state." Ever the student of detail, Pearsall described it as "412 square miles of mountain ridges and hollows [in a] dendritic pattern."[36]

Mobility in a place like this was not easy; the accumulated effort struck Pearsall as more than just overcoming obstacles. Mountain mobility practices were a social factor for which anthropologists should account when they tried to impart a sense of place. In this view, places were always unstable relationships between lived landscapes. Paths of circulation were performed by daily movements for work, for errands, for visiting, and for the sheer joy of going anywhere. Some of Pearsall's most striking narratives told of local trips in cars, trucks, and jeeps. She filled her notes with descriptions of the hours she spent in transit. Local roads were often half-completed or, as she put it, "narrow, unfinished, unbridged, and usually unscraped." Pearsall drove from Hazard to Buckhorn one Sunday in June 1959. Of the 24 miles she covered, the first 15 were paved and the last 9 were gravel. This Perry County trip took her through "wild and quite isolated" country. On the other end of the county, people living along Jacks Creek were effectively cut off from Hyden for most of the year. When they needed to acquire license plates, take a driving test, pay county taxes, or get a TB test, they rattled their way to the county seat. In dry summer months, the trip was a 19-mile journey over rough roads with a 3-mile unimproved section. During the rest of the year, Jacks Creek residents had to complete a staggering 75-mile trip to get to Hyden, via Beverly (on a road that was completed in 1955), Pineville, Straight Creek, and State Route 421. Hyden's Dr. Bowling rarely made house calls to Jacks Creek, or any hollows, for that matter, owing to the state of creek roads.[37]

Mobility in Leslie County was more than an arithmetic of distances, travel times, and costs; it was also a social practice that reinforced and reshaped the relationships at play in overlapping neighborhoods. Automobility, as scholars currently use the term, denotes the political processes that govern movement, the cultural arrays that legitimize movement, and the ways of experiencing the world via motion. Automobility in Leslie County was a good proxy for the host of problems that came with entrenched poverty, limited investment in infrastructure, and an aversion to anything that smacked of socialism. Many people who lived in eastern Kentucky spent their entire lives embedded in the "local" networks of family and daily subsistence. Others migrated permanently or cyclically within new employment and kin networks outside the mountains. Both groups relied on the "political-geographic positioning" that established social chances for individuals. Pearsall engaged in automobility research decades before it became a pressing concern for social scientists.[38]

Poverty produced a marked hierarchy of funding, access, and disruption. One paved road had served Leslie County from 1931 to 1945: the 20-mile route from Hyden to Hazard. Only after World War II did the state pave the 34-mile route south out of Hyden to Harlan and the 29 miles west to Manchester. Both additions were mining routes, a trend that continued well into the 1970s. The county's roads were busy, even before the era of mass car ownership. The seemingly unending circulation of vehicles on inadequate roads was caused by the prerogatives of the mining industry. The Berea College economist David Brooks called eastern Kentucky the "home of the small, locally owned 'truck' mine" in 1967, and it had taken the better part of two decades to win that title. Getting coal and timber out of the mountains competed with getting tourists into the mountains as economic planners' top priorities in the 1950s and 1960s. Drivers hauled all Leslie County coal to the railroad links at Manchester and Hazard. Owning and operating a coal truck could make a man a relative fortune. Pearsall learned of the Couch Brothers on Bob Fork, who had multiple coal trucks running in 1959 and seemed to be thriving. During Pearsall's residence, coal trucks rolled through Hyden constantly. They roared as they moved through town shortly after dawn and on until late evening. When she accompanied locals on the best roads, they always passed a steady stream of coal trucks. Pearsall counted over thirty trucks on one trip between Manchester and Hyden. When they had the choice, residents avoided roads that were used heavily by coal truck drivers. Better to take one's chances on narrow, rutted trails than to meet a coal truck careening around a bend.[39]

In terms of Leslie County automobility, then, mining companies controlled the ebb and flow of traffic. The Leatherwood mining complex, just across the Perry County line to the east, was connected by the Louisville & Nashville Railroad to the towns of Hazard, Krypton, Chavies, and, eventually, Lexington. The road from Hyden to Leatherwood was 14 miles of paved road, followed by 10 miles of "good dirt road." Leatherwood was connected to Hazard by a paved highway entering from the east. Traveling from the west, the pavement began at the Perry County line. Positioned between Cutshin Creek to the west and Leatherwood Creek to the east, the mining town was a threshold between the relatively developed corridor of western Perry County and the less developed eastern fringe of Leslie County. To Pearsall, communities along Cutshin Creek looked "relatively prosperous." She found "fairly broad and level" land along the road that paralleled the creek, which meant some tobacco and more corn than she had seen elsewhere. "Good houses," she noted, "all apparently with electricity." All of this had come from mining-related investment in infrastructure.[40]

Even though extraction industries' needs governed the road system, locals made it theirs. In July, Pearsall noted in her field book, "People around here are certainly not untraveled but they travel to work and they see and live in the working class world wherever they go." The road from Hyden to Harlan, wrote Pearsall, "twists and winds through narrow valley floors along the Middle Fork with steep sided ridges on either side and a number of breaks between ridges with roads going back to various neighborhoods." Hundreds of people lived within a mile of the road, and when the state converted it to a "good blacktop highway," it reoriented the residents of the county's southern half. Pearsall reckoned that they must have associated themselves in the 1930s with Hyden and the "lower" (northern) part of the county. Now, they were in Harlan's orbit, despite the journey's steep climb over Pine Mountain. An accessible route to Harlan meant that its Miners Memorial Hospital now attracted more patients, leading to the closure of smaller institutions like the Pine Mountain Hospital. From the perspective of people who lived in lower Leslie County in the 1920s, there was no way in or out of the southern half. The road changed that, and it was the coal economy that brought the road.[41]

People who knew the roads knew how to make do with their piecemeal nature; Pearsall had to catch up quickly, because she could not avoid leaving the county's 77 miles of blacktop in her travels. Her journey from Hyden to the Red Bird Mission, where she consulted nurses, was 7 miles to Stinnett on paved roads and then 12 miles on dirt roads "through a series of narrow and twisting valleys." On this middle stretch, Pearsall observed "practically nothing" by way of settlement. She bounced through the trees along Stinnett Creek and Woodson Fork, a passenger in a jeep operated by the FNS. As they approached the Red Bird River, she began to see a few sawmills and houses. Construction was occurring on a road to connect Bowen Creek to Jacks Creek, but Pearsall's driver kept to the rocky Bowen Creek road. It ended at a paved county highway, and the Red Bird Mission was another 6 miles to the south. The trip illustrated how a little insider knowledge helped. The majority of the drive went through what looked like the middle of nowhere. Without signposts or clear maps, the wild intersections offered few hints for orientation. Half-hidden side roads that seemed too small to be through routes turned out to be the only ways forward. In addition to the chaotic pathfinding, drivers had to cope with the conditions of the paths. A resident joked to Pearsall that there were only three seasons locally: snow, mud, and dust. Bad roads turned terrible by bad weather could make many of the county's cars useless for days on end. During a particularly wet week in the spring of 1960, Pearsall met a family "coming over the mountain" near Long Branch. Two mules carried two young boys, suitcases, and sacks, while parents and an older sibling walked. During and after heavy rainstorms, Pearsall wrote in her field notes, "much walking along all the roads."[42]

Drivers tended to be male breadwinners; thus, most members of a given family could be stuck without wheels for large portions of the workweek. Nevertheless, the county buzzed with a variety of traffic. Produce was sold from trucks that crept along the back roads, X-rays were taken from a "truckmobile" driven by a husband-and-wife team from Lexington, two pickups served as a rabies clinic along a 30-mile route, the county "bookmobile" forded streams as it made its way up hollows, and a "gospel team" truck started rolling as soon as the winter snows ended. On a very rough stretch of road along Hell for Certain Creek, Pearsall passed a "good pickup truck," a 1950-model Chevrolet, and a decades-old Model A Ford. "None of the cars and trucks were really late models," Pearsall wrote, "but most were only a few years old and apparently in fairly good condition." On Saturdays in particular, there were so many cars in Hyden that people parked them three deep in front of the courthouse. Drivers who found their cars blocked in had to make a circuit of town, looking for the owners. Many drivers left their keys in their cars to expedite the process. The mobility offered by cars brought people into town to socialize more than to shop. Pearsall found stores rather empty on weekends, despite the busy streets.[43]

A crucial means of transportation for those without cars were the clinic jeeps, which the FNS staff used to pick up the sick or injured and transport nurses around the county for house calls. Jeeps could ford creeks that stopped even pickup trucks with a running start. Staff at the health clinic in Hyden regularly heard from patients who could not make it in because the "water was up" on their branches. The Cutshin Clinic staff also had a jeep that served the southeastern part of the county. FNS nurses picked people up whenever they saw them walking along roads, whether they needed medical care or not. When Pearsall accompanied an FNS nurse on a trip to Trace Fork, they picked up and dropped off fourteen people in a day. Most families up the creek had trucks or "fairly good cars," wrote Pearsall, but the fords were so bad that using the jeep when it passed was a welcome opportunity.[44]

Locals also used a communal car system that was organized by creeks and fueled by patience. On upper Phillips Fork, for example, one car and one truck served all residents as needed. Pearsall frequently saw people sitting by the side of the road waiting for a ride as she passed the Phillips Fork turnoff.

This could be a prearranged system with neighbors or an ad hoc occurrence, as when a girl living on Big Laurel Creek was badly burned and her family carried her out to the road and hitched a ride to the Cutshin Clinic. Hitchhiking was universal, as Pearsall discovered early in her stay. As she left Confluence to return to Hyden in the second week of her fieldwork, a boy flagged her down to ask if she could transport him, his three sisters, and his mother. The trip to and from Hyden was the most popular in the county. People arrived in groups on Commodity Days and waited around for each other to complete their business. The process of getting into town was so demanding that even when people showed up on the wrong day for commodities, county officials obliged in order to save them the trouble. School-bound children often started out on foot to a main road, rode in a passing car with other kids from nearby creeks to a bus stop, and then bused to school.[45]

For those without a functional car or a cooperative neighbor, mining companies ran buses around Leslie County to coincide with shift changes. Area private bus service began in the 1940s and connected coal towns with people living up the most populous creeks. Morning, noon, and evening buses served Hyden when Pearsall was in town, with the bench in front of the town's drugstore serving as the main stop. One bus line connected Hyden and Hazard, another ran west to Manchester, and a third line ran along the Middle Fork from Confluence and then "up Cutshin" to Leatherwood. The last carried miners who lived along Trace Fork, just above Confluence, to the Blue Diamond mines. Pearsall noted that this route was possible because the road to Confluence was so good, owing to extensive lumbering that had been done along Trace Fork. Dr. Bowling said that each morning brought a "rush of people . . . from around the county by bus," who all tried to catch the noon bus back out of town. Like the roaming jeeps, the county's buses provided residents with a routine network of mobility.[46]

Finally, taxis were popular in the immediate vicinity of Hyden, if one could enlist fellow passengers to split the tab. A taxi driver in Leslie County could be a professional operating out of Hyden, Harlan, or Hazard, or simply a resident with a car who charged a fare. Sharing a taxi with one or more people often cost the same as a bus ride. The cost of a taxi was unregulated and varied greatly, but Pearsall gathered what details she could. A county resident who needed to get to the Cutshin Clinic paid between $1.50 and $2. (The mining bus "up Cutshin" cost $0.83 each way.) A taxi to the Oneida Hospital in Manchester cost as much as $13. One Leslie County family paid $15 for an "ambulance taxi" to the Miners Memorial Hospital in Hazard—

and $18 for the return trip. A one-day round-trip to Louisville with the right driver, on the other hand, might cost $20 to $22. A two-day chartered taxi trip to Louisville cost the Baker family $50 in early 1960. Meanwhile, bus fare from Hazard to Louisville was "a little over $10." Taxis were not the last resort for people who needed medical care and had no other means of mobility. The most difficult cases for public health leaders, Pearsall observed, were those for whom ambulances were the only option. For these people, the remoteness of their homes or their antagonism for kin meant that health care was largely self-administered. If an ambulance driver had to rush from Hazard or Manchester to the houses along Camp Creek or Long Branch, it was certainly an emergency that resulted in a daunting bill.[47]

All of this activity relied on working vehicles, and Pearsall found that owning a car and maintaining it were two very different things. "Sure, the [Baker] family has a car," she wrote, "but the tires are badly worn, the transmission is about to go, and they can't afford gas; these sorts of things can be said about many of the car owning families." The run-down state of cars in the area became a dire complication when residents relied on them for medical care. Pearsall heard of multiple families using their cars to make repeated 30-mile and 100-mile round-trips over rough roads, rather than leave a loved one overnight in a hospital. When those cars would not start or broke an axle, two weeks of treatment for anemia or long-term rehabilitation for a twisted back ground to a halt. There was a less serious side to the makeshift state of cars, too. Pearsall noted that few vehicles in the area had mufflers; she heard the traffic outside her window from the moment she awoke. Buzzing, throbbing cars could be heard from miles away as they worked their way up or down twisting creek beds. Fixing cars was as ubiquitous an activity as driving them. Eddie's Garage in Hyden was one the area's premier socializing spots for men. Jim Woods ran a general store down on Bullskin Creek, and Pearsall saw eight old cars for sale outside. Junk cars were everywhere, waiting to give up useful parts. Local cars were in such bad shape because the condition of the roads destroyed them, and residents' income levels pushed them to try partial fixes rather than full replacements.[48]

For Pearsall, the final connection between automobility and healthways accompanied a screeching of tires and a shattering of glass. Auto accidents occurred at the intersection of inadequate infrastructure, frenetic industrial rhythms, and a social system that valued specific types of independence. Although Pearsall's field notes can make it seem as if cars were crashing every five minutes, it was a result of her sharp focus on mobility and health care access. Of the sixty-three people who died in Leslie County in 1959, half of them died from heart attacks, strokes, and cancer; only five people died after any type of accident that year. Of those five fatal accidents, three of them were car crashes. Much more common in local newspapers and Pearsall's notes were head-on collisions (or near misses) in which drivers and passengers broke bones, smashed their heads, cut themselves on broken windshields, and bruised their ribs. The bad conditions of roads, the limited sight distance in narrow valleys, and widespread drinking meant that some type of incident on the roads occurred weekly. "There seem to be a good many automobile accidents here," Pearsall wrote, "and Rufus says most of them seem to be young people who are driving recklessly around mountain curves." One afternoon while she was climbing out of Greasy Creek, she came across two cars that had just crashed into each other on a blind curve. It was one of thousands of such spots in Leslie County, where drivers temporarily lost sight of the road and hoped for the best. The crashed cars had been filled with people, all of whom were "shaken up and bruised and cut." Pearsall marveled at what happened next: they pushed the cars to the side of the road, hitched rides with drivers who had appeared on the scene, and went about their business. The next morning, as she typed her notes, Pearsall remarked, "No one seemed concerned about the injuries, and no aid was sought."[49]

The crash above Greasy Creek came several days after Pearsall watched a car roll off the road into a creek 50 yards below. Car accidents became almost physical presences in the social life of the county. Pearsall referred to a man who lived "down the road from Miss Tolk," who was paralyzed from the waist down after a crash in 1958. She passed a young couple one night who were waiting on the steps of the Hyden doctor's office, faces caked with blood, having been dropped off by a state trooper after they had driven into a tree. When she spent mornings in the town café, customers reported on the fate of those who had crashed overnight. She heard of multiple victims: the Asher boy recuperating in a Louisville hospital, the man who died in the Hazard hospital, the man with a broken back who hitched a ride home from the Cutshin Clinic after "violently and successfully demanding his release," and the coal truck driver who begged onlookers to shoot him as he was dragged from his fiery cab.[50]

From talking to locals, Pearsall found two schools of thought. One held that the crashes were caused by youthful indiscretion. After the sheriff left town at 9:00 each night, a nurse named Betty Roberts told Pearsall, the "young fellows" began to drag race along Hyden's two streets. Those races turned into chases out in the steep hollows, and someone often paid the price. The other school of thought understood car wrecks as a symptom of an ungovernable world. Those who died or became gravely disabled from accidents received great sympathy and concern from the people with whom Pearsall spoke. And those who survived with minor injuries received praise for dodging their fate. The key was that drivers were seen as independent operators—lone agents pitted against the uncaring universe or in cahoots with an accommodating God. Their kin, their unwitting passengers—all of those people mattered less in these narratives than the central figure who did not slow down, because slowing down would do no good. Some survivors seemed almost legendary, like the man from Bullskin Creek who put $100 down on a car on a Friday and wrecked it that Sunday while driving drunk. Pearsall noted that he had moved through "a succession of three or four cars" since the first accident, and locals talked about his audacity. He kept escaping, and it seemed like he might keep on escaping.[51]

To Pearsall, the car crashes of Leslie County made the most sense when viewed through the framework of change in a small community. Car crashes brought the physical, social, and cultural frontiers of Leslie County contact together in a way that was immediately productive for the researcher. She understood that preaching to residents about safer driving was ineffective, and it seemed that building better roads might only encourage greater speeds. Easy solutions were not to be had. But the skilled behavioral researcher who took such matters seriously would see that "healthways" and "roadways" were manifestations of the same layered, fluid social relations that included residents, nonresidents, and itinerants like herself.[52]

Sociologist John Stephenson noted that he gained credibility during his fieldwork in the mountains of North Carolina when he flipped his car in a ditch on a narrow road. "News of this accident made the rounds of the community like lightning," he wrote, "and my name and reputation were established." Pearsall did not make her name by crashing a car. If there were a name to be made locally, it would be through nonthreatening "snooping around" and research findings that improved the interactions of residents and medical personnel. Externally, she would make her name through establishing a methodology that opened up the mountains.[53]

The Difficult-to-Place Stranger

In January 1960, as Pearsall was organizing her field notes and attending to administrative duties in Lexington, *Harper's* published Wayne Kernodle's

"Last of the Rugged Individualists." Kernodle, a sociologist at the College of William and Mary, warned that highway construction was destroying the isolation that had produced cultural difference. His case study of "Pin Hook Gap," North Carolina, highlighted the "glory and the sorrow" of road development and economic change. Among the aspects of life that Kernodle thought had become standardized in the mountains in the 1950s were manners, clothing, wages, and speech patterns. He noticed civic and social clubs replacing the informal gatherings that had organized social life for generations. Progress brought loss, Kernodle stressed, because Americans were running low on heroes. When everyone lived like everyone else, it took extraordinary effort to find inspiration. Kernodle's article epitomized the way in which nonresidents' observations of the mountains reinforced the very phenomena that they purported to reveal.[54]

Pearsall was not worried about the fate of mountain residents at the hands of "progress." Change had always been a part of their lives, and lamenting the passing of a timeless cultural type was misguided. In the face of poverty and logistical extremes, the people she observed crafted routines for themselves which revealed the ubiquity of change in their daily experiences. It was mistaken to brand them as "stagnant," but it was also incorrect to claim that they were in the midst of unprecedented flux. Both viewpoints missed the intricacies of how people lived now and long ago. Greater penetration and accessibility in the mountains did cause new problems of identity, but perhaps more so for social scientists than for "mountaineers."

Pearsall based her mountain expertise in part on her ability to access the inaccessible and translate the rhythms of life in Leslie County. Her summer 1959 work was devoted to the "more accessible parts" of Leslie County; she planned to move into the "more isolated sections" to the west in 1960. After several days in Jacks Creek and Phillips Fork in the early spring of 1960, she scoped the communities along Cutshin Creek in the southeastern corner of the county. While she was there, she developed pneumonia, which (along with the "pressure of work" at the University of Kentucky) cut short her plans for several more months of research in Leslie County. The aborted fieldwork also seems to have scuttled her plans for a "healthways" book.[55]

In Pearsall's wake, the rest of the nation climbed up into the mountains for fact-finding tours. John F. Kennedy visited West Virginia as part of his primary campaign that spring. A year later, when announcing the passage of the Area Redevelopment Act, Kennedy referred specifically to the "problems" of eastern Kentucky as a cause for national concern. The media coverage that followed his campaign focus on Appalachia reinvigorated the "rediscovery" of Appalachian poverty by federal agencies, regional planners, and social scientists which had been building for years. For a generation or two, Leslie County had been a place where outsiders could gain experience. The proprietor of Pearsall's hotel said that plenty of young women came to the area over the summer to work as FNS couriers. The couriers, she noted, were "rich girls who come here for a vacation or to have a new experience." The "Back of Beyond" image lingered, even for those who were adamant that Appalachia was neither extraordinary nor alien.[56]

In the immediate years after Pearsall's stay in Leslie County, two other published works focused researchers' attention on the difficulties of implementing social change in the mountains. In 1962 the University Press of Kentucky published *The Southern Appalachian Region: A Survey*. Edited by sociologist Thomas Ford and with a foreword by Berea College president W. D. Weatherford, the book attempted quantitative analysis that rejected determinism. What had been made could be remade, the authors concluded. The survey attempted to de-exoticize Appalachian residents, reminding readers that what they assumed to be distinctive mountain traits were actually rural traits that could be found across the country. The scholarship presented within was action oriented and forward looking. Southern sociology icon Rupert Vance ended the book with a prediction about southern Appalachia in the 1990s. By then, the "lost generation" of mountaineers would be dying off, and their lack of economic productivity would go with them. Meanwhile, decades of job training, education, out-migration, and "family limitation" would have integrated younger generations into the national economy. The mountain culture of maladaptation and stubbornness was but an episode in the broader history of the American mainstream. Ford's survey voiced a kind of wide-eyed optimism that a combination of intervention and time's passage would shift Appalachian foundations significantly.[57]

Then, in 1965, Jack Weller published *Yesterday's People*, based on a decade and a half of social work in the remote mountains of West Virginia. Weller took a slightly different approach than Ford's contributors, finding less adaptability and more pathology in the residents of Appalachia. Weller's title summed up his interpretation of the people with whom he worked: they had been "caught for generations behind the wall of hills" and looked back nostalgically to a fantasy world of independence. Change could still occur, Weller stressed, but it would take much more effort than Ford's relatively confident volume suggested. Lest the book promote defeatism, Rupert Vance was again on board to frame Weller's outlook. Vance referred to a "new regionalism" sweeping American social science in the early 1960s, an impulse that echoed the motivations of action research. "Since mountains are not likely to be moved," Vance wrote in the foreword to *Yesterday's People*, researchers were beginning to devote themselves to methodologically sound intervention. Difficult work lay ahead.[58]

Meanwhile, Pearsall was moving forward with close analysis of her research process, even if her ground-level work in Leslie County had ended. She argued in 1965 that behavioral scientists must mount a focused review of participant observation, because it was rarely clear how researchers were actually using it on-site. Fundamentally, she wondered whether the participant observer's role, when introduced to social groups, influenced scholars' data. How could one measure the effect of "mutually meaningful relations," especially the scenario in which the researcher became changed by the research process itself and "behaved" the change while interacting with group members? If nothing was static, then the measurement tools had to change. The degree to which the visitor participated mattered immensely. If a scholar adopted the "observer-as-participant" role, then he or she tried to stay aloof enough to win residents' candor—assuming that they told an "attentive stranger" things that they would not disclose to more intimate relations. If a scholar adopted the "participant-as-observer" role, then he or she hoped to gain special access, a social and cultural immersion that cost time and energy. Pearsall acknowledged that the researcher in the field bounced between these positions, all the while dealing with insiders' reactions to the "difficult-to-place stranger."[59]

A social worker from West Virginia University captured the challenge after working on a 1964 project with the rural poor of the Mountain State: "We need some completely new methods of working with people, and I'm not sure what they are myself." Pearsall played with what Columbia sociologist Robert Merton called the "stranger value" of being an outsider. Cornell sociologist Harrison Trice likewise encouraged researchers to cultivate their "outsideness" tactically, being careful not to misalign themselves within the "honeycomb [of] face-to-face informal groups and congeniality groupings." Pearsall cited both Merton and Trice as exemplars of a willingness to adopt and shed roles as necessary to get a deeper sense of how members interacted within social groups. "Outsideness" could be manipulated in such a way by skilled researchers if they recognized that the barrier between residents and nonresidents was real but contingent. Pearsall's instructions for fieldwork started with advance study of the site for the purpose of identifying the best entry point. Next came establishing the legitimacy of the project with official and unofficial leaders, followed by a "getting-acquainted period in which the observer is himself observed." The researcher should adopt a visible activity, she advised, like making a local map or chart that required him or her to visit and talk with group members. This gave the researcher a pretext to circulate and be seen. Next came the slow work of planned observation to record "the necessary evidence for generalizations about who and how many behave in stated ways under specified conditions." Interviews followed and overlapped with long-term observation. Pearsall noted that interviewers should always consider the validity of their subjects' information, especially their motives for confiding in the first place. Field notes had to be comprehensive and resourceful. Every day accumulated. Everywhere shifted. Everyone was someone else's difficult-to-place stranger.[60]

Pearsall did not have a magic wand to fix Leslie County's healthways, nor did she claim that swift progress was likely. She wrote in 1962, "In the long run, of course, success depends on local acceptance of a new set of health beliefs and practices which satisfactorily replaces old ways in what amounts to a thorough-going re-education process. This in turn, to close the somewhat pessimistic circle, is almost inconceivable in the absence of major economic and sociocultural change." The lack of quick fixes was not surprising; the anthropologist could not change decades of practice after six months of fieldwork. What she could offer were concrete steps for other field-workers. Rendering technical reduced Leslie County to a set of conditions—a field of professional practice—but Pearsall asserted that there was nothing reductionist about her attention to social life in the mountains. If others learned from her the ability to see hidden flows of identity and communication, then the generalizing was worth it.[61]

Pearsall shared Weller's goal of figuring out the rules and rhythms of the place and "put[ting] it down for the help of others who would work in the mountains." Answering colleagues' call for systematic commentary on the "human relations skills which go with the social researcher's role," she published four articles on the nature of the Leslie County challenge in the next five years. During that period, she helped found the Society for Medical Anthropology and the University of Kentucky's Center for Developmental Change. By 1964, she held joint appointments in the Department of Anthropology and in the College of Medicine's Department of Behavioral Science. An editorship for the journal *Human Organization* soon followed. Professionally, the 1960s were Pearsall's most productive and wide-ranging part of her career, with scholarship ranging from a comprehensive bibliography of medical behavioral science to an examination of sub-Saharan African development. If she had made other scholars notice her potential as a field researcher in Berkeley and Alabama, her work in eastern Kentucky had confirmed that she was an authority on field techniques—at holding a mirror up to social scientists as they plied their trade.[62]

Mary Breckinridge had been right, in a way, about the county possessing the "priceless asset" of isolation. Certainly, reality and image were separated by a gulf that was easier for some to believe in than for others. The isolation of the mountains was a paradigm that had to be discredited through the force of will before it stopped motivating research. There was still a cachet to fieldwork on distant ridges and up dark hollows; anthropologists and sociologists were introspective enough to recognize it, but that did not mean that they were immune to it. By focusing on the methodology of Appalachian participant observation, Pearsall revealed the performance behind intervention. The hardships of Leslie County were many, but for the skilled investigator, the isolation of the most remote hollows became a compelling part of a résumé.

William Gedney and the Look of Coal Country

The photographs were framed on the living room wall of a house in Oklahoma City, 900 miles from the mountains of eastern Kentucky and twenty-five years after they were taken. On the left was a portrait of a woman and her daughter, both looking down, arms tenderly wrapped around each other, the woman's cheek nestled against the child's temple. The scene looked like a stolen moment—that instant when all was quiet in their world and the embrace meant everything to both of them. They were Vivian Cornett and her two-year-old daughter Bernice, sitting on their front porch in Big Rock, Kentucky, in July 1964. The photo on the right showed a middle-aged man in a filthy, torn white T-shirt. Arms folded in front of him, hair slicked to the side, Willie Cornett looked like he might have been smiling a moment before the shutter opened. He, too, looked down at the ground, in a pose that acknowledged the presence of the photographer but did not address it. He leaned against a windowsill outside the same house. These were family photos, taken by a man who arrived abruptly in the summer of 1964 and was gone in no time. He mailed the prints to Vivian several months later. This sort of exchange had been occurring in the mountains for decades. Traveling photographers had worked the back roads of Appalachia since the turn of the century, taking pictures of anyone who would pay for their printing. Cameras came into the mountains on foot, on horseback, by mail order, and by car.[1]

But William Gedney, the man who took these pictures, was not an itinerant "camera man" looking for clients; he was an art photographer from New York, looking for a career in the mold of Walker Evans and Robert Frank. He wanted to shoot an unvarnished version of American life and display it for people who were ready to receive it. The fact that Wilma Jean Cornett, the daughter of Willie and Vivian, still hung these pictures in the late 1980s, when she had left the family homestead in Leslie County, was a testament to Gedney's skill at capturing ephemeral moments of humanness.[2]

In July 1964, Gedney spent a month in the hills of eastern Kentucky, living with the families of two coal miners not far from where Marion Pearsall had worked four and a half years earlier. Gedney, who was thirty-one and looking for a project to launch him into the world of his idols, used dozens of rolls of film as he observed the residents of Leslie and Perry Counties preparing meals, improvising games, fixing cars, and resting. He returned eight years later, by which time his career had taken him across the nation as a Guggenheim fellow and to India for a year on a Fulbright Fellowship. Gedney took hundreds more Kentucky photographs in 1972, many of the same people in the same types of everyday practices as he had captured on the first trip. The training that Gedney gained in Kentucky was in the sensitive, controlled portrayal of people who reminded viewers of figures from photojournalists' spreads in the early 1960s. Gedney's subjects *looked* like the Appalachian poor from *Life* or *Time* or the *New York Times*, even if they did not fit into the mountain melodramas of the era. He attempted to avoid generalizing and stereotyping his subjects, but the success of his photos relied on the recognition of the Cornetts and their neighbors as types. Supporters praised his images from Kentucky as an artist's response to mass media clichés—here was proof of what a talented photographer could do to find complexity in scenes that had become formulaic. His 1964 portfolio won him the praise of key tastemakers in New York, especially the new photography curator at the Museum of Modern Art (MOMA). Gedney's achievement was to merge the formal analysis of the physical surfaces of eastern Kentucky, the stories of work and "home turf" which fascinated him throughout his career, and the cultural symbols of 1960s Appalachia.

Behind it all, and even in front of it all, was the landscape of coal. Gedney wrote four pages of notes in July 1964, when he had just arrived in Kentucky. He was already thinking of how he might explain to future critics the photographs he had not yet taken. He set the scene to capture his first impressions of the Cornett house, editing himself as he wrote:

> The family of Willie Cornett live two miles in a holler in eastern Kentucky. The mountains in eastern Kentucky ~~are not like the ranges in upper New York State~~ come together ~~down~~ in sharp V shaped valleys these are not broad lush valleys of farm land with carefully tended houses surrounded by car well mended fences

> the mountains come close together, the vegetation is thick and rough infested with snakes. At the bottom of the V runs a road, unpaved, twisting through creek beds (over makeshift log bridges). Full of large holes and boulders ~~the mountains coming close together the road following the base~~ people live along this road. Sometimes the road (if it can be called that) connects after 5 miles or more with main road more often it just goes back as far as it can go and stops. They call these hollers. Willie Cornett, his wife Vivian and their 12 children live two miles back in a holler.

The cyclical run of the description suggests that Gedney wanted to translate his project to external reviewers. When he reached the point of placing the Cornetts in an eastern Kentucky hollow, he realized that the local topography would have to be explained to anyone not from the southern Appalachians. So he provided context, trying to convey the feeling of being in the mountains, particularly on the roads that meandered between them. But he did not adopt the view of so many visitors, who looked out their windows as "narrow, bumpy paved roads dissolve[d] into muddy trails, connecting the Appalachian South's bleak hollows with the world beyond." To make meaning out of the images, one had to consider the way in, through tight, winding valleys on roads that made every mile a headache. But "bleak" was a matter of focus and frame.[3]

This type of place setting was fundamental to Gedney's photography. Many of his Kentucky images placed the lives of his subjects in local landscapes. Gedney was not a landscape photographer; nor, he noted, was he a "social problem" photographer. He was primarily a street photographer dedicated to images in which form and content merged. This made the mountains of eastern Kentucky particularly attractive. He took pictures along the border of geography, autobiography, and metaphor, the trio of references that the photographer Robert Adams argues is central to noteworthy landscape photography. The geographic aspects of Gedney's images related to the ways in which isolation was broken down by patterns of circulation. He was fascinated by the materiality of Appalachian Kentucky and the place-making qualities of his subjects' built environment. Autobiography came into his images through his preoccupation with his grandparents' rural rhythms. He caught echoes of their farming lives through his pictures of people in mountain hollows. Finally, the metaphorical aspects of his photographs revolved around the ubiquity of cars in the lives of the Cornett family. Gedney came to see the automobile as a primary symbol of American life, and the effort that his subjects expended to keep their cars running seemed like a larger commentary on their marginality.

The Problem Crescent

Gedney had been working for less than a decade in photography when he started the Kentucky project. He graduated from high school in Upstate New York and then completed a graphic design degree at the Pratt Institute. His post-Pratt work in magazine production and design (for Condé Nast and Time Inc.) and in advertising (for Watson Products) in the late fifties and early sixties seemed to be a stopgap solution to the expense of living in New York City. Gedney cultivated his photography and turned to it exclusively in 1964, when he had the money to quit his commercial work. Even before that, he produced images that revealed his fascination with everyday scenes, the built environment, and people at work. He shot deliverymen around his neighborhood in Brooklyn. He shot his grandparents and their upstate farm in 1956 and 1958. He shot dozens of pictures at a Tribeca wholesale market in July 1962. A theme that preoccupied him from the fifties to the seventies was the world that people made for themselves day by day. He approached his subjects on their own terms, while producing photos that did not aspire to biography. Viewers might learn nothing about the specific people depicted, but they could sense moods and moments. There was an anonymity to his imagery which avoided sentimentality.

Gedney arrived in Kentucky with the name of a contact, possibly given to him by a journalist who had made a similar trip from New York. He waited in Hazard in the last days of June, hoping to get approval from a local family to board him for a week. His first break came from Boyd Couch, a forty-nine-year-old miner whose family lived in house 169 of Blue Diamond Holler in Leatherwood. Couch was the recording secretary for the United Mine Workers (UMW) Local 8280. Everyday life in Leatherwood had taken a dramatic turn three months before, when the Blue Diamond Coal Company shut down operations and announced that it would no longer negotiate with the UMW, which had represented Blue Diamond miners for two decades. Disruptions like this, as well as a general downturn in the coal industry, were part of the reason why 75 percent of those between the ages of fifteen and thirty-five left the mountains in the twenty years after World War II for manufacturing jobs in cities like Cincinnati and Detroit. Some commentators, like the minister Jack Weller, argued that the migrants were the "strong, young, and ambitious ones," leaving behind the "psychologically immobile." The youth of the migrants was undeniable; their psychological superiority, a matter of perspective. With only one out of six young residents of eastern Kentucky finishing high school in the late 1950s, entry-level work in industry offered the best chance at income. But that, too, came and went. Out-migration ricocheted when economic prospects dwindled in industrial centers. Chad Drake, the managing editor of *Mountain Life and Work*, observed in 1958 that layoffs in Ohio and Detroit led to two unexpected turns of events in eastern Kentucky: an influx of the recently departed and an increase in petitions for public assistance. Between 1960 and 1965, the percentage of Leslie County's labor force employed in mining dropped from 48 percent to 28 percent.[4]

The southeastern corner of Kentucky had attracted more attention in the four years since Pearsall had "snooped around"—and much of it had involved mining politics. National journalists had taken new interest in the area's economy after Kennedy's 1960 primary tour. With headlines like "Portrait of an Underdeveloped Country" and "Valley of Poverty," news articles on Appalachian coal miners appeared steadily between 1960 and 1965. In the late summer of 1962, Perry County became a media focal point when wildcat strikes, roving pickets, and a wave of arson met mine operators' refusal to pay into the UMW's welfare fund (and the UMW's subsequent revoking of area miners' health care benefits). This was a rehash of the violence that had accompanied UMW organizing campaigns a decade before. The landmarks were still there, though Gedney would not have known them. A rutted driveway on Bob's Fork was where Bleve Begley's car was blown to bits in 1952. The Hyden-to-Manchester road gained fame in 1953 as the route along which a UMW organizer was murdered as he sat in his parked car. In 1962, a bombed church, a damaged railroad bridge, and wrecked coal tipples became the latest signs of the labor–management fight.[5]

Gedney's experience in Kentucky was quite domestic. The Couches had six kids, three of whom still lived at home. "The family is all," Gedney noted on July 5; "it is the unit of society." His images consistently captured domestic situations, either the everyday minutiae of large nuclear families or the comings and goings of extended families, as relatives and longtime friends socialized. This contrasted sharply with the overview of the *New York Times*'s Homer Bigart, who noted the previous autumn that Leslie County alone reported one hundred desertion cases a year. The hills were filled with broken families, the press reported, driven by grim calculations. Public assistance policies favored abandoned wives and children, prompting men to disappear, at least when welfare officials came to investigate. The writer Harry

Caudill told Bigart that residents had redubbed one Leslie County settlement "Widowsville." In Perry County, nearly 40 percent of residents qualified for federal assistance.[6]

Gedney stayed with the Couches until the middle of the month, when he moved a few miles to the west to stay with the Cornetts. The move from Leatherwood to Big Rock was a step away from the relatively good infrastructure of the mining camp and toward the less developed zone of Leslie County. A pair of economists studied eastern Kentucky's industry and infrastructure and identified a "problem crescent" created by mountain ridges and creeks. The problem could be measured in low rates of telephone access per capita or in miles of paved roads, but Mary Jean Bowman and Warren Haynes were more intrigued by the way that fertility rates indicated topographical isolation. The neighborhoods perched on the highest ridges or wedged into the most remote hollows exhibited fertility rates drastically higher than the rest of the state. Even as these rates slowly decreased following World War II, they remained noticeable outliers.[7]

The Cornetts lived in one of these remote corners of the "problem crescent," 2 miles from a county road and the creek that paralleled it. Roads and creeks—infrastructure and nature—were the focal points of much activity exerted in the mountains by outsiders in the sixties. Fifteen months before Gedney's trip, record flooding hit the area after two weeks of heavy rain. The damage in eastern Kentucky and West Virginia prompted the Kennedy administration to establish the President's Appalachian Regional Commission in April 1963. A year later, Lyndon Johnson touched down on a poverty fact-finding tour. Johnson visited eastern Kentucky two months before Gedney, arriving in a helicopter from Huntingdon, West Virginia. Press photographers snapped hundreds of images of Appalachian porches that week, especially when Johnson stopped to talk with residents around Inez. The Fletcher family lived 5 miles out of town and packed ten people into a small house that Johnson reached by leaping over a runoff ditch at the side of State Route 3. National correspondents detailed the "rude front porch" of the "tarpaper shack" and its "grassless, hilly yard . . . filled with junk" as an odd place for a president to be, but one with rhetorical power as the site to launch a war on poverty. When a wind gust blew over the door of an outhouse in the Fletchers' yard, the Associated Press was there to cover it. The University of Kentucky's Thomas Ford counted thirty-nine grant-in-aid programs established to address eastern Kentucky's deficiencies between 1961 and 1966. Aid agencies perceived most of those deficiencies as economic in nature, dealing with what Bigart had called a “vast ghetto of unemployables” nine months before Gedney’s arrival.[8]

The coverage of poverty and workers’ strife generated a steady stream of private relief efforts as well. The phenomenon was epitomized by Alma Moy, a student at Barnard College. In late 1963, Moy and several dozen other students from northern universities caravanned to eastern Kentucky to distribute food, clothing, and money to poor families. Crammed into cars and accompanying a “truckload of warm clothing” on the nineteen-hour drive, the students had learned of the economic conditions in the coal district from news reports and a campus speaker. They received organizational support from a local branch of the Retail, Wholesale, and Department Store Union and coordinated with counterparts in Philadelphia. Interviewed by an Associated Press correspondent near Hazard, Moy remarked that most of her party had never “come into contact with a situation like this before.”[9]

The chances that outsiders would learn of the Appalachian “situation” increased, owing in large part to two CBS documentaries broadcast in 1964. In February of that year, Walter Cronkite took his *Twentieth Century* show to Manchester for an installment titled “Depressed Area, U.S.A.” Dirt roads, junk cars, and pronouncements about “despair” focused viewers on the grim scene of unemployment and poverty. The show’s writer, Earle Luby, had issued a summary report from his scouting trip the previous summer, and the visuals corresponded to his initial impressions. Luby wrote of “old, abandoned automobiles strewn like beer cans along the sides of the roads by the hundreds and perhaps thousands. Men and boys crawl over them seeking spare parts for those machines which still run.” Luby saw in the people and the ubiquitous junk the “nadir of human life and pride in America.” Willard Van Dyke’s direction took the viewer alongside Cronkite as he visited families beset with disability and malnourishment.[10]

Charles Kuralt followed in December 1964, with a tour of Letcher County for his *Christmas in Appalachia* program. Kuralt presented the tedium of unemployment and the environmental ruin in which a million people “waited.” *Christmas in Appalachia* began as so many accounts of Appalachia did in the 1960s—on a road that the visitor hesitated to call a road at all. Kuralt and cameraman Walter Dombrow took viewers off the road to interview residents in their scattered houses, including a “strong woman, surrounded by exhausted men, and living in an exhausted land.” The Kuralt broadcast prompted donations from people and organizations around the country, funneled through CBS and national and regional charities. An anonymous donor from Florida paid for a Christmas party for three hundred area kids, with gifts of shoes, clothing, and 150 hams.[11]

The imagery that came out of eastern Kentucky in the several years before Gedney's visit was what he called "social problem" photography. It was photography that relied greatly on captions to deliver its message. It was through captions, for instance, that readers of a January 1964 *Life* article on Letcher and Pike Counties knew that two children looking curiously at John Dominus's camera were "ragged urchins." It was through captions that readers knew that two men in blazers and hats looking to the right of the camera were "standing in the watery sunlight, without work and without purpose." Finally, it was through captions that three figures above a snowy waterway became a family that "trudge[d] across a rickety suspension bridge over a sewage-polluted creek to its two-room shack." Weller remarked on travelers' inability to see any beauty or value in Appalachia's visual scene. Instead of the "well-kept garden patches" and "picturesque footbridges" that dotted Lincoln County, West Virginia, visitors' attention was drawn to peeling paint, the rough blocks that raised houses off the ground, the porch rocking chairs and the people who rocked them, and the "dumps of trash spread everywhere." News photography led viewers down narrow interpretive paths, ending at a tale of victimization or local color.[12]

Gedney did not set out to counter these approaches; he wanted to extend his own approach. His Kentucky work showed clear parallels to his first project, completed six years earlier. This was an effort he called *The Farm*, a series of photographs that he made 30 miles southwest of Albany, New York, at the home of his maternal grandparents, Ezra and Mercedes Tryon. He took the bulk of the pictures in 1958 and shot both interiors and outdoor scenes on the 111-acre site. In terms of technique, the Norton Hill photographs relied on the shifting play of sunlight in buildings and exterior spaces and trained Gedney in capturing subtle light conditions. His notes documented rooms in the farmhouse "lit by natural light, never by artificial light" and the experience of being in the house during the day's "gradual failing." There was an elemental quality to his stay at the farm, and his photographs represented an attempt to render the experience of being removed from city life. Several years later, he revived this goal in Kentucky, where he shot frequently in shadowy interiors during the evening and the misty countryside in the morning.[13]

Gedney liked rural rhythms. He marveled at the "full feeling of a day encompassed by rise and setting," as shadows moved slowly throughout the farmhouse. His memories of childhood trips blended with the sedate atmosphere of an elderly couple on a farm that was winding down, making his experience of the place in the late 1950s a feeling of comfort and routine found in worn surfaces. He had tactile memories of the site—the cold stone walls, the rough beams in the hay loft, a ladder's smooth rungs, and the dust floating in the glow that forced its way through gaps in barn doors. He explored such memories for their photographic potential. Scenes played out in his mind. The interior of the farm's largest barn was the epicenter, with light cutting in from all angles, yet "even in brightest sun there were parts of the barn that remained dark." The barn, like the outdoors, was "mysterious" and "vast" to Gedney, both as a child at play and as an adult with a camera.[14]

Gedney's grandfather died in 1961, and his grandmother died a year later. After they were gone, he compared their lives to those he saw around him on the streets of Brooklyn: "I cannot imagine my grandfather awakened by an alarm clock, gulping down a cup of coffee, a piece of toast, rushing to the office that nervous queasy feeling in his stomach. Their lives had dignity and meaning. Their lives were balanced well ordered." Elsewhere, he referred to their lives as "integrated," with work and leisure barely separated and without reference to the increasingly pervasive radio or television. The barn and the surrounding fields acted as a "requiem" for his grandfather. The farmhouse did the same for his grandmother.[15]

Two direct influences on his Kentucky work flowed from his time in Norton Hill. First, Gedney discovered that some people accommodated themselves with their environment more than others; they adapted to their conditions over the years and became at ease in indifferent landscapes. He was convinced that his grandparents had learned the secret of being happy in routines that would drain others. His impressionistic writing searched for the feeling of the farm: "I find a harmony in my grandparents' lives. a peaceful settlement with reality that few of us achieve. their surroundings and their existence are bound inseparable—the warm dark smell of the barn . . . lingered about my grandfather. there was a feeling of wholeness nothing held back figures in the landscape as if the earth had given them birth like a tree. they were rooted in the land." His grandparents were embedded in massive sets, the barn and the fields "not conforming to one person, size, or need, but for a larger scale or need." Wet grass and dusty hay left their marks on Gedney and the Tryons. Dirt and gravel were everywhere. "My grandmother would hear our car coming up the lane," he reminded himself in a notebook, "and she would come to the door as we parked by the house." Everything seemed to fit, and the apparent stoicism of his grandparents made Gedney pursue photographically the human ability to be in place.

Second, the *Farm* project taught Gedney how to notice arrays of materials, how to fit them in the frame, and how to use inanimate objects to suggest decades of recurring activity. He wrote, "Rooms settled. No new furniture appears here. New material objects are not required by (old) age. The furnishings are comfortable, familiar objects, whose touch has become part of those who have shared their existence. Like clothes that take the form of bodies after years. Objects that have served you become part of you." The Tryons' homemade furniture impressed him with its signs of age. The old floors creaked under footsteps and rocking chairs.[16]

As he photographed and reflected on the Tryon Farm, Gedney developed an idea that geographer Tim Cresswell has described as "doxic landscapes." These are landscapes of practice, in which people's daily routines and preconscious movements solidify into accustomed experiences of the world. Doxic landscapes are the "ruts" that people wear into their everyday environment. Gedney saw such ruts everywhere. The house's rooms, he wrote, were "formed almost by their steps," and the house itself became the "area surrounding their steps." Photographs of empty rooms could speak to the decades of living and working in that spot. The myriad tasks that his grandmother completed each day became "patterns worn in the floor in certain tracts [*sic*] that also must exist in the air that fills the room." The "simple, strict order of objects give beauty," Gedney wrote, and what might appear as sparsely furnished or cluttered rooms represented something much larger for him. Doxic landscapes could be captured on film, in what he later called "non human views." *The Farm* taught him the value of such views. He might not have expected the glut of junk that he encountered in Kentucky in 1964, but he was prepared to portray it.

"Neither More or Less"

Despite the context of unemployment, resentment toward both unions and employers, and casually accepted violence, but aware that it had become eastern Kentucky's calling card, Gedney resisted framing his subjects and their environment with the tropes of economic desperation and ecological exploitation. He indexed the material minutiae of the sites to reflect on specific relationships between people and places. The raw materials of his images, then, were roughly two dozen people, a handful of houses built in hollows, yards and porches that became work sites and play sites, and an accumulation of discarded items (cars, car parts, house wares, furniture, bottles, boxes, boards, etc.). A photograph taken in such a setting could end up showing too much. Gedney's task was to manage the clutter.[17]

He believed that his best images captured "in a single frame visual forms organized to the point where neither more or less are needed." What he sought was the "single moment when form and content become one." The clutter of eastern Kentucky could be read as either form or content, but finding a way to present them as a coherent and original statement about people and environment was not easy. Photographing "junkscapes" could lend itself to the spectacular mode of photojournalism, with the viewer asked to marvel at how pathetic people managed to live in such circumstances. Gedney framed junk in a way that avoided condemning lifestyles or praising perseverance. This was the stuff of life. As Geoff Dyer put it, "If there is a wrench in Gedney's photographs, that is because it is going to be picked up and used to mend something, to fix something, to repair something." Everything was up for grabs. One reviewer noted that the children in his Kentucky images seemed not to notice their surroundings, whereas the adults had resigned themselves to it. Routines slowly wore ruts throughout these homesteads. His images portray local cultures as so local that they frustrate attempts to use them to make claims about Appalachia as a whole. Doxic landscapes are site specific; the form of built structures and pathways reveals content about individuals' daily engagements. Gedney later grouped such images under the heading "Details of American Life," pictures "that together will form a portrait of our culture."[18]

A pair of images from his 1964 trip displayed this "neither more or less" approach to local scenes. First was an image shot in the second week of July in Leatherwood (fig. 5.1). The photograph shows a woman turned away from the camera and leaning into the driver's window of a 1951 Chevrolet Business Coupe. The woman and the car occupy the middle ground of a vertical composition. Above her is a densely forested mountainside. Below her is a small swath of junk, with scrap metal, plastic, and tires strewn across the foreground. The composition is careful and quite playful. Gedney labored to fit multiple, textured layers into a single frame to depict a relationship that emerged as he walked around the car. At first both the woman and one of the passengers saw Gedney shooting the car from behind. They noticed that he was taking pictures and stared directly at him. Unless he was shooting formal portraits, however, he did not make many images in which his subjects addressed the camera. He preferred a sense of spontaneity, so he walked to his

5.1. William Gedney, Untitled, 1964 (woman looking in car window). William Gedney Photographs and Papers, David M. Rubenstein Rare Book & Manuscript Library, Duke University.

left to stand behind an intervening strip of auto parts. The woman at the car window went back to her conversation, and Gedney kept moving behind her. The final print showed Gedney fashioning a vision of Leatherwood from a mountainside of trees, a thirteen-year-old car, an elderly woman, and a patch of junk that was as thick as he could make it.[19]

It was the type of scene that he photographed often during his ramblings around Brooklyn, but what set this apart was what sits above and below—or before and behind—the car. Many photographers chronicled the environmental degradation of Appalachia through deforestation and mining. Gedney rendered what was typically depicted as decay or violence into a *strategic* clutter. "Garbage disposal," Gedney wrote in his notebook, and the phrase contained a world of its own. Almost every outdoor image he made showed garbage disposal, in the sense that junk was everywhere. Weller wrote of West Virginia families using their yards and creeks to hold "heaps of stinking refuse, old cans, bottles, bedsprings, and cartons." It might have been hyperbole to suggest that a trip into the hollows in the summer "was enough to discourage the strongest stomachs," but Weller knew, too, that weak stomachs were a sign of unfamiliarity. Gedney attempted to bridge the familiar and the unfamiliar. Leatherwood was cluttered with heaps of wood, heaps of cars, heaps of tin cans and glass bottles, heaps of cinder blocks and broken appliances. It fascinated him that the heaps were not truly discarded, but were constantly scavenged and reworked. Residents carved out space for themselves in the hills and modified that space constantly. If they lived amid litter, it was because they did not equate it with disorder. People and things circulated in the Appalachian valleys, and Gedney wanted to convey that circulation—the kinesthetics of the place—to the viewer. A photo like this could portray cultural practices without making them worthy of ridicule. He moved to a position that made the collection of junk large enough to suggest that it was not incidental; this was how people chose to live.[20]

The materiality of daily routines—what geographer Nigel Thrift has called "the little, the messy, and the jerry-rigged"—comes through in this picture. Making numerous exposures of the ephemera of his subjects' lives, Gedney documented the look of a place shaped by distinctive ideas about people, possessions, and land. Together, these themes offered a way to dramatize the ordinary, to offer a study of Leatherwood which conveyed his fascination with the *stuff* piled up around the site. That pile of junk accumulated years of residents' specific activities and efforts, like the worn furniture in his grandparents' farmhouse. This was not Appalachia, the photo urged, but rather one small fraction of Leatherwood. This was not "the usual disorder of coal country," to quote Bigart in 1963, but rather Boyd Couch's neighborhood. Gedney conveyed a sense of people making do. They had figured out how to live in places that demanded ingenuity.[21]

A second photo showed Boyd Couch at the home of his father-in-law in Grassy Branch Holler (fig. 5.2). When Gedney moved to the location 10 miles away (requiring a drive of more than 25 miles), he shot scenes of hillside farms perched above thickly wooded creek beds. His vantage point and his

5.2. William Gedney, Untitled, 1964 ("Boyd on hill"). William Gedney Photographs and Papers, David M. Rubenstein Rare Book & Manuscript Library, Duke University.

human subjects were above the valley. In Grassy Branch Holler slopes led away from Gedney to the unseen branch. It was a subtle shift in perspective from Leatherwood, but it opened up the hollows of eastern Kentucky in a way that many photojournalists could not achieve. Photographers' journeys to communities in mountain hollows often ended where the road ended, at the base of hollows. Images taken from these depths made homesteads and residents appear suffocated by brush or half-hidden by trees.

In this photo, which Gedney referred to as "Boyd on hill," Boyd appeared below the camera, turned down the hill. Boyd stood between the house and a shed, and the lower half of his body was obscured by a row of pants drying on a line. One could miss Boyd at first, amid the landscape below him, a series of overlapping slopes that dipped into the hollow. Three birdhouses atop thin, leaning tree trunks, two houses, and small fields of corn receded into the photo's background (or Boyd's foreground). Gedney captured what Weller called the "assorted" look that came from layouts that were more functional in the moment than carefully planned. The image was a study in belonging in place, contemplating how people became part of a locality. The Couches and the Cornetts could trace their ancestors back to the earliest white settlers along Cutshin Creek and Clover Fork in Leslie and Perry Counties. They seemed to be a part of their surroundings in a way that struck Gedney as profound.[22]

This shot could be read as Gedney's response to the many photographs of bleak hillside farms perched above thickly wooded creek beds, a visual shorthand for Appalachian misery in 1964. The picture belied a passage that Gedney copied in his notebook about Kentucky, quoted from the epilogue of Caudill's *Night Comes to the Cumberlands*: "Exhaustion is apparent on every hand—exhaustion of soil, exhaustion of men, exhaustion of hopes." The shot of Boyd above the valley conveyed activity instead of exhaustion. Lives were being lived here, and the topographical and material piling up testified to an abundance of practice. There was potential, not decay, in this image. Boyd was not hemmed in by the debris around him; he had navigated it for decades and had worked it into his patterns. If one could not quite see the patterns of routine movement in the air, as Gedney imagined, one could at least sense the patterns from the ease with which his subjects existed in the space.[23]

He chose both of these images as potential covers to a book on eastern Kentucky. In February 1973, as he worked on a dummy for the book, he alternated between these two as iconic images to summarize the project. Thinking of these as "landscape scenes" suggests how Gedney saw landscape as an assemblage of people, terrains, and built forms. He wrote about the "polluted water and useless soil" of eastern Kentucky, but his images treated ecological concerns as less interesting than the feel of specific places—the moods created by the depth and busyness of compositions. When he made these editorial decisions in 1973, it was in the aftermath of another major project that was influenced by and focused his framing of eastern Kentucky.

This second project was his sixteen-month Fulbright Fellowship in India. Gedney arrived in Delhi in November 1969 and started shooting in the oldest parts of the city. He quickly seized upon the "feeling of confusion and clutter, the closeness of human beings," as the primary mood for his India images. His relatively brief time in Delhi was a prelude to a longer residence in the country's most sacred city, Varanasi. When he moved there for a year-long stay in early 1970, he continued to grapple with photography's relationship to places that he experienced as "the surfeit motion, the noise, as if the streets and the passage in the streets were endless." A major Hindu pilgrimage site on the Ganges River, Varanasi was a spectacle of poverty and religious devotion for Gedney. It forced him to put into context the jumbled living he had photographed six years earlier in Perry and Leslie Counties.[24]

Outdoor work was a daily occurrence in every street of Varanasi. Gedney took pictures of street performers, barbers, rickshaw drivers, laundrymen, furniture makers, cooks, weavers, ditchdiggers, musicians, and street vendors. He photographed people bathing in the Ganges, sleeping on doorsteps, performing religious rituals, cremating loved ones, and walking casually down streets flooded with shin-deep water. "Life in an Asian street is truly life," he observed, "meaning all human activity, the essence of existence is carried on in the streets." There was undoubtedly chaos in the Varanasi experience, but it contained rhythms that Gedney slowly discerned. Over time, he discovered "order in the streets . . . certain patterns not at first apparent." Drivers seemed to go anywhere they pleased, but they never seemed to hit pedestrians, cyclists, or sleepers. Vendors spread themselves out all over the city, yet their stalls formed makeshift corridors and squares that came and went with predictable regularity. Some of the worst breakdowns in the life of the city were the result of government officials trying to alter practices that worked according to a "flexible, rough logic." Gedney noted that the clogged traffic around Varanasi's central traffic circle was created by the installation of new traffic lights, a pet project of "progressive" administrators who wanted to make the city more efficient. He disdained such improvement campaigns.[25]

Despite the scale of the place, despite the population density around every corner, Gedney took photographs to show that life in Varanasi was conducted on a "very personal level between human beings." Each day seemed to mirror the one before, but the people he observed on the streets subsisted with apparent patience and grace. The photographer struggled to find new subjects and angles by which to shoot them. Gedney even complained to himself about his diary, noting that the book was a "dull recapitulation of the days' 'events.' " The model of the philosopher Eric Hoffer's diary, published just the year before, was surely on his mind as he assessed his own journal. Hoffer had blended the minutiae of working as a longshoreman with the insights of a voracious reader and thinker. Gedney had met Hoffer several years before in San Francisco, and a relationship of mutual admiration and support had ensued. Yet Gedney could not do for Varanasi what Hoffer had done for the docks of San Francisco. Instead, he used his diary to help make sense of his photographs. There was so much activity, so much life in India, yet the experience became generic after weeks and months. The diary became "an attempt to keep track of the days which slip bye [*sic*] so easily here in India, a watch on myself."[26]

Photos of Indian streets challenged viewers in the United States to think about their own everyday practices, and this challenge had arisen in Gedney's work during his Kentucky trip. He did not want to celebrate or condone the lifestyles he found in these places; rather, he looked to present them as some entryway into larger aspects of American life in an era when everything seemed disposable. The daily work that he saw everywhere in India made him think again of his Leatherwood and Big Rock pictures. The public evidence of people's possessions suggested an alternative process of accumulation which operated in both places, despite poverty. Gedney's time in India made him think seriously about producing a book from his Kentucky photographs, and he made his second Appalachian trip a year after returning from his Fulbright Fellowship.[27]

An example of his India–Kentucky connection is a photo of Willie Cornett working on a car in the front yard of his house (fig. 5.3). Cars were a consistent presence in Gedney's career. In 1971 he toyed with the idea of a book of automobile images, given the "scattered" material that he had produced over the years. Kentucky provided many. The "Appalachian junk car" was a staple of 1960s journalism, with writers tallying the number of broken-down cars they passed during their travels and using the objects as metaphors for the state of mountain people. The "hardy roadside blossom of eastern Kentucky," as the *New Yorker*'s Calvin Trillin put it, suggested stagnation. However, more often than not, the cars depicted in Gedney's photos were being actively worked on. As he had on his grandparents' farm, he tried to aestheticize the performance of this labor. His work images in Kentucky were not of mining, but of preparing family meals, moving furniture, and, above all, repairing cars. Cars meant so much to the Cornett men that it struck Gedney as fitting that the oldest son, Junior, eventually got married in a gas station in Sneedville, Tennessee. The way Gedney heard the story, "Junior and his girl Linda were so hot/anxious" to wed, that the most convenient option was an auto mechanic turned justice of the peace who would not leave his garage but was still willing to officiate. "It seems appropriate," Gedney observed, "since so much of the thoughts and actions of the male Cornetts involves automobiles."[28]

As he walked around the Cornett house one afternoon, he came across Willie and several of his sons working on a mid-1950s Cadillac that they had pushed onto a ramp. As he often did, he started by taking close shots of the work, moving in as the Cornetts secured the car and ramp before attempting to fix the engine. Kentucky homesteads reminded Gedney of the surfaces

5.3. William Gedney, Untitled, 1964 (man working on car). William Gedney Photographs and Papers, David M. Rubenstein Rare Book & Manuscript Library, Duke University.

that fascinated him in Norton Hill. Everything was basic, without ostentation or elaboration. In Leatherwood and Big Rock, however, the simplicity of homemade rooms and furniture displayed a lack of "finished craftsmanship," an incompleteness that betrayed residents' priorities. They did not care for the look of things in general, and they certainly did not fuss over the interior of their houses. The wear of their furniture, like the state of their yards, told the tale of years' worth of use.[29]

Gedney's preliminary exposures focused on Willie and his sons problem-solving, with Gedney as attentive to their faces and tasks as they were to the car ramp. The exposure that Gedney selected for printing set the work within the context of the Big Rock homestead. He thought of almost too many meanings for cars in eastern Kentucky; it was hard to pin any of them down. Cars served as altars, gathering places, and wombs; they were there to "be leaned against to be felt, touched"; they were tactile, leaving "the taste of gas in mouth oil on body, grease on hands and face." They were forces of destruction, hurtling down bad roads and crashing at spots that drivers pointed out to passengers, "like awards." Cars dominated local conversations, they were in a perpetual cycle of breakdown and recovery, and they symbolized both mountain masculinity and mountain futility. Finally, cars were a way to tell time, as local residents dated their memories by the car they were driving at particular points in time.[30]

The image followed Gedney's tendency to decenter the human figure, placing Willie and the car amid a yard dotted with car parts, empty bottles, a table with an ashtray, weeds, and small plots of crops. This was public work—as public as any street scene in Varanasi—yet the photographer was the only spectator. The Cornett house stood to the right, with a child and hanging laundry on the front porch. A hillside covered in trees extended past the top border of the photo. Behind Gedney, as he took the picture, was a wide clearing formed by the dirt driveway into the hollow and a makeshift parking lot holding several cars. Gedney framed the image to omit that sense of openness. Instead, the work area, the house, and the surrounding hills were stacked on one another, in a style reminiscent of Walker Evans's 1935 photo of a steel mill, houses, and a cemetery in Bethlehem, Pennsylvania. Although Gedney's photo did not share the vista of "Boyd on hill," the viewer was again called upon to view a scene as if one were spying on other people's lives. The turn away from the camera here comes across as occupation—Willie was too busy to pose with the car, too engaged in fixing it to act the part of the Kentucky miner with the broken-down automobile. The image forced a double take: once as a straight shot of a physical scene, and twice as an allusive image of toil which echoed New Deal murals.[31]

A fourth photograph, from Gedney's last days in Big Rock, portrayed the young generation of Cornetts who had not been trapped (yet) in regional ruts (fig. 5.4). For some external observers, the sight of "abandoned cabins up narrow hollow roads" was the telltale sign of migration forcing residents into the future. Thomas Ford and his colleagues with the Southern Appalachian Studies project concluded that the region was overpopulated with people who would slowly redistribute themselves to find work. For Gedney, the top transformative force was the creep of new roads, and the way they allowed towns to "sprawl out" even in the deep forest of the mountains. The younger generation could "escape to the city," Gedney noticed in 1972, but they still returned to their homes at the end of the weekend. The uneven economic development represented by the new roads brought a "supermarket sameness" to eastern Kentucky, which Gedney loathed but the youth of Leslie and Perry Counties appreciated.[32]

5.4. William Gedney, Untitled, 1964 (children walking down road). William Gedney Photographs and Papers, David M. Rubenstein Rare Book & Manuscript Library, Duke University.

He was slightly puzzled by what he called the "mystery of personality" on display among the Cornett children, kids in 1964 and young adults eight years later. The centripetal force of family encouraged kids to fall in line, yet symbolic displays of youthful revolt erupted frequently. Trouble was one of the main ways in which children drew attention to themselves, as with what Gedney called the "revolt of Ken," who was fourteen years old in 1972. Still, Gedney noted a "lack of rebels in [the] family to break the pattern." Toward the end of his 1964 stay, he accompanied a group of Willie and Vivian's children as they walked from their home to Creed Cornett's. Gedney lagged behind to photograph four Cornett boys walking along a rocky dirt driveway to the small house. Four abandoned cars lined the path, and the brush to either side was littered with broken fences and nondescript debris. Two children waited on the house's porch for the others to catch up.[33]

The house became the Cornett children's playground as Gedney photographed them. Subsequent shots showed them posing for Gedney behind the wheel of a car, doing handstands in the yard, and then climbing onto the roof of the house. The photo that Gedney selected from the batch showed none of this frivolity, and it could certainly be assumed from this image alone that the children all lived under that roof. This was as close as Gedney came to mimicking what he called "social-problem photography." But he saw it as a product of different motives. When he reflected on such images in the early 1970s, he explained it this way: "I am concerned with making . . . an uncropped blending of form, value, and content. I prefer the ordinary action, the intimate gesture, an image whose form is an instinctive reaction to the material. In these photographs I hope that something of the region and its people is conveyed. . . . Reality is elusive and this is only my view."[34]

The photo was an overt way to rehash the isolation trope that appeared so frequently in media reports. Gedney believed that family life in Kentucky hollows was a reaction to small clusters of people being left to their own devices. He thought he learned a lot about interpersonal relations during his six weeks on the ground in 1964 and 1972. The family profile he read into his images was of families spinning in cycles of tension and cohesion, bound by their size and environment. The kids walking together, but apart, exemplified the vagaries of cohesion. The Cornett family was simply too large, and their corner of Leslie County too small, for placid relations. No one could count on privacy. None of the children could expect to wear clothes that had not been passed down through their siblings. "Group ownership," Gedney wrote, "never having anything of your own." Family members loved each other, he knew, but they frequently fought over frustrations that brought violence to the surface. The Cornett and Couch parents raised their children with "scattered discipline," though they did not seem to have the time or the energy to devote attention or "individual care and love" to them. Love was group centered, as were most aspects of daily life. Sleeping arrangements bred closeness, as Gedney observed that "sleeping together many to a bed caus[es] a clinging to each other in life." The hollows of Kentucky were not isolated, but life within families like this was "self-contained, self-absorbed, insulated."[35]

It is hard to know just what Gedney thought of the people of eastern Kentucky. He described them in his notebooks with such phrases as "food stamps and medical debilitate initiative," "male dominance–female slave," "lack cleanliness—baths and sex," and "anti black feelings." That bleak portrait must be tempered, however, with the possibility that Gedney listed both the observations culled from his time in Kentucky and the nationalized narratives about Appalachia which he studied after 1964. He knew that he was not breaking news about the mountain coal region, yet he learned details that made him reassess what it meant to live there. They were not easy lives to live. Houses featured hanging photos of dead children displayed in their coffins. Gedney visited Willie's father's house, where he learned of the violence that haunted the family. Two of Willie's brothers had fought while drunk a year earlier, and one of them had died. "Ten children without a father," Gedney wrote in his notes. In the main room of the house, he saw a large photograph of the ill-fated brothers with their arms around each other. He thought that they must have been teenagers in the photo. "Eyes at the camera . . . faces frank, joyful, youth and its dreams. How many years ago was that?" Gedney puzzled.[36]

He later learned that Willie's generation had been left to fend for itself. Willie's father had frequently left them in a car while he drank in the bars that catered to miners leaving work. Gedney cringed at the thought: "three or four in the morning. They waiting. Forgotten. Tired asleep." As they left Grandpa Cornett's house, Gedney studied a portrait of a fourteen-year-old girl ("she looked about twenty"). It was the last image of Willie's sister, who ran away and got married very young. Gedney's notes suggest that he got the full story from Willie or his father: "five weeks after she was found with her head blown off by a shotgun. (Her husband at work, the stove going, but ice in her hair.)" Elsewhere, Gedney wrote of the "suicide attempt of Jr." and the history of past violence that most adult men seemed to carry with them.

If he avoided sentimentality in his pictures, then he also avoided the urge to present anything about eastern Kentucky as simple. Nostalgia, he wrote, was the "security blanket of the present," and he did not think that a photographer should betray the richness of everyday life in particular places, at particular times. Between the farm lanes of Upstate New York and the packed streets of Varanasi, the hollows of eastern Kentucky helped launch Gedney into the urban art elite. Though set in a landscape that many depicted as spectacular, Gedney produced images that were defiantly ordinary. They portrayed people who, he explained, were caught in "mental and physical depression" with a "complete lack of future and hope." But rather than look the part of broken mountain folk—the people who appeared in magazines and on television screens—they went about their daily routines with a casual, if not joyous, air. When Gedney made images of these people in their environment, he hoped to deliver a visceral blow to the viewer, not through the pathos of poverty spectacles but through the tactility of the site as translated by film. Through offering his "sure, thoughtful embrace of a piece of the physical world" in each shot, he argued that the people and things of Big Rock and Leatherwood were adaptable, challenging the established narrative of such places as static and failed.[37]

Untouchable and Uncommunicable

It is in the reception of his work that we can see Gedney's achievement of an expertise revered by certain art photographers of the 1960s. In particular, the Kentucky project was key to advancing his career in the New York art world. Margaret Sartor notes that Gedney had only a handful of true friends in the profession, although his private notebooks show that he networked with plenty of photographers and editors in the 1960s and 1970s. His career's traction took hold shortly after his first trip to Kentucky. The output from that month spent in the mountains, shared with a few key gatekeepers in 1965 and again in 1968, changed his trajectory significantly. The interpretation of his Appalachian work placed him in the context of a street photography cohort that pursued multiple ways of depicting the richness of everyday life, without, as the curator Lisa Hostetler argues, "preconceived notions of value and certainty." Although he never gained the fame of contemporaries such as Diane Arbus, Robert Frank, or Danny Lyon, Gedney wedged himself into their milieu through his pictures of the Couches and Cornetts. We can trace his emergence through four developments in the decade after he began sharing these pictures: two honors bestowed on him from leading institutions of contemporary art and two dead-end attempts to seize upon that increased scrutiny.[38]

First came the John Simon Guggenheim Memorial Foundation Fellowship, announced in April 1966. The Kentucky images, along with pictures from the Tryon Farm and portraits of classical composers he took in 1965, composed his application portfolio. Without the stamp of the mountains to give his portfolio a topical focus—*Appalachia*, at a time when it was a vivid concept and a hot media commodity—it is hard to imagine him receiving the Guggenheim Fellowship. Of the past decade's worth of photography fellows, ten had been photographers of social landscapes (mostly urban streets), three had focused on natural landscapes, two had pursued more experimental forms of study, and one had turned a photojournalist's eye to the civil rights movement. Gedney's innovation was to take the preoccupations of street photography and apply them to the locales of eastern Kentucky. The result was to convey a sense of "found scenes" in places that took a significant amount of effort to reach and inhabit. Letters of recommendation from the legendary

Walker Evans and the MOMA photography director John Szarkowski certainly helped Gedney's chances. Gedney knew Evans from Time Inc., where he had also allied himself with influential picture editor Ben Schultz. It is unclear when and how Gedney and Szarkowski became acquainted, but it is likely that Gedney submitted pictures (perhaps from *The Farm*) for review before going to Kentucky. Evans and Szarkowski endorsed Gedney's photos as important statements of creativity and talent. Both men were former Guggenheim fellows, Szarkowski most recently in 1962, right before he assumed the directorship at MOMA. Their status as fellows mattered, for the award process could be a partisan affair in the 1960s. Of the seven men on the 1966 review committee, only one, the painter Robert Motherwell, had not held a Guggenheim Fellowship. The committee's choice of Gedney, along with fellow American-centric street photographers Arbus and David Vestal, marked the older generation's validation of an inquisitive yet removed style of photography by people who were generally ten to twenty years younger. Along with the architecture and landscape photographer Paul Caponigro, Gedney was the youngest of the new Guggenheim fellows, at thirty-three years old. Within a year of his return from Big Rock, then, Gedney had an impressive team behind him.[39]

The Guggenheim Foundation funded Gedney for a project couched as "Photographic Studies of American Life." His ability to complete such work was measured by his portfolio, especially the pictures that, he wrote, came from "a series taken in eastern Kentucky on the people of the depressed coal mining areas." By providing such context, Gedney made his case for the deep meaning that lay within images of everyday beauty. His intended meaning was an authenticity that came from impermanence. He caught everyday scenes that were tied to a specific set of conditions—mountain poverty, political impotence, technological lag, and so on—that were not replicated just anywhere. For those who vouched for his artistic merit, Gedney's pictures perhaps provided what the scholar Ardis Cameron has called "places of authenticity and cultural intensity seemingly absent from the home they called 'America.'" Authenticity is a slippery term, to be sure, one that says more about the appraisers than the objects appraised. The intensity of the culture that was looking back at viewers from his pictures was tied to his written description: not just "people," but "people of the depressed coal mining areas." Anyone who encountered American mass media in the mid-1960s would have known what that meant, what it suggested about the residents depicted. Gedney might have been trying for a perfect confluence of form and content, but the "social problem" context of his work was necessary to build resonance.[40]

Like expertise, authenticity is a "quality of a process rather than an object." When applied to the visual arts, it is a way of saying that the artist did it—produced the picture—the right way. The images coming out of Kentucky seemed authentic because Gedney's approach passed the test of observational, "found" photography. In this view, he was a street photographer who had the courage to take his lens to places that were simultaneously unlikely and familiar. A key word in the professional reception of Gedney's images was "unobtrusive." It was a term that Szarkowski would later use when introducing Gedney's work to the public. Framing the Appalachian series as "unobtrusive" highlighted the lack of interference in his approach and confirmed the authenticity of his process; Gedney did it right. His pictures were evidence of time spent on the ground, watching, learning, and not pushing wantonly into the lives of his subjects. Unobtrusive images came from biding one's time, until the photographer and the camera blended into the background of the subjects' daily routines. Noticing without being noticed was crucial to this aesthetic, which put Gedney at odds with the professional news photographers who developed Appalachian poverty stories in the 1960s. He composed his prints in reaction to what seemed like the sensationalistic, interventionist treatments of print media. His Guggenheim application played up the authenticity quotient by explaining further that the collection's underlying theme was "the human being in conflict with, and in relation to, his environment."[41]

Szarkowski was a tireless advocate for street photography techniques and their ability to present something new about the world. When he took over as the MOMA photography director, he told *New York Times* photography writer Jacob Deschins that the "only way to discover a new form is to know so precisely what you think of the subject-matter you are dealing with that, when you see the result, you know that the photograph is an accurate echo of your own feeling when you took the picture." Gedney never wrote at length about his thoughts on the subject matter of eastern Kentucky. It is likely that he never quite decided what the subject matter actually was. If, as he told the Guggenheim Foundation reviewers, the subject was the relationship between people and their environment, then Gedney, like Pearsall before him, saw multiple environments affecting the people of Leslie and Perry Counties. But he could certainly relay detail that he collected during his stay. He gained insider-style information. He learned that for people with little exposure to television, *The Munsters* was the most recognized pop culture reference. He learned that children went to the bathroom in groups. He learned that everyone could tell the difference between individual cars, as the noise of their engines echoed through the trees. He learned that adults had an alert system for approaching figures—a scratch on the left side of the nose meant a woman was coming, whereas a scratch on the right announced an incoming man. This type of detail came through in his pictures as an emphasis on how people in Leslie and Perry Counties *did* things—how they fixed cars, walked to neighbors' houses, plucked chickens, cut hair, slept, and whiled away the time.[42]

Szarkowski's own photography had focused most recently on landscapes in the Quetico-Superior Wilderness of northern Minnesota and southern Ontario. He especially appreciated pictures that avoided condescension and moralizing. His support of the most recent Guggenheim photography fellows was zealous. When he selected the pictures of Guggenheim winners for MOMA exhibitions, he signaled what raised them into the upper echelons of esteem. He praised images that were "more distanced, less willful, seemingly more objective" (in reference to Lee Friedlander, 1960). He applauded "unembellished" observations, "so free of apparent craft or artifice" (Bruce Davidson, 1961). He liked photos that conveyed "stability, simplicity, and great emotional force" (Dave Heath, 1963), but he also sought rebels whose "taste for life" trumped their "regard for art" (Garry Winogrand, 1964). Above all, Szarkowski wanted to experience the "honesty of vision" (Diane Arbus, 1963) that drew the "grace, drama, humor, pathos, and surprise" out of the "immemorially routine acts of life" (Helen Levitt, 1960). This group of photographers, despite their sometimes-contentious relationships with each other, subscribed to an aesthetic in which "tough" images (those that were "stark, spare, hard, demanding," in Joel Meyerowitz's terms) were also beautiful images. Szarkowski's support suggests that he saw some of these qualities in the eastern Kentucky pictures.[43]

Gedney used the $5,000 in Guggenheim funding to support a cross-country road trip that kept him moving until February 1967. By August, he was shooting pictures in Chicago and Iowa. He moved through the Mountain West in September, before arriving in San Francisco in October. For four months, he photographed people and places in the Bay Area, Los Angeles, and San Diego. He experimented with night photography and associated with people on the fringes. Eric Hoffer, the laborer turned Berkeley adjunct, was a strong influence on Gedney at this point. Gedney quoted at length from

Hoffer's *Working and Thinking on the Waterfront*. A passage that stood out was Hoffer's observation that to be civilized is "to rise above passion; to be able to observe and report without giving way to anger or enthusiasm and let the (viewer) reader react as he will." If Hoffer thought that the definition had value, it was certainly true for Gedney. He longed for photography that conveyed what he considered straightforward messages—*this is what it was like*. Above Hoffer's remark in his memo book, Gedney wrote "Walker Evans?" The documentary aesthetic had worked in the mountains because it was still possible to believe that not all that could be said about these people and places had been said. Evans had modeled a matter-of-fact approach to the world which gained a new currency in the hands of the younger photographers whom he supported.[44]

The attempt to rise further above passion brought Gedney to a second significant milestone: the creation of a book to collect his best work to date. Evans had released two books of photographs (his first in decades) the previous year, and Gedney's admiration for his quasi-mentor may have moved him toward trying his hand at publication. Evans's 1966 books focused on hidden-camera subway photos (*Many Are Called*) and a career's worth of interior shots (*Message from the Interior*), both suggesting deep, complicated lives that were led far from the photographer's lens. In the subway series, nineteen of the eighty-nine pictures featured passengers who stared back at the camera, reminding the viewer of the intrusion that the photographer committed. Amid the series of a dozen interiors, only two included people within the frame; the other ten "pretend[ed] to reproduce," in Szarkowski's phrase, "the very bones and clay of the actual world." Evans's two collections offered evidence of established patterns of living, both in the individual lives that were suggested and in the accumulation of lives that encountered specific spaces. And they testified to the photographer's stoicism, what the photography curator Colin Westerbeck calls indifference transformed into impassivity "in the presence of strong feeling." Szarkowski praised the *Message from the Interior* images as appearing "without interpretation, without feeling, almost without thought." Gedney followed suit in his selection of pictures. He strove for an unassuming air in his book project, which he worked on in 1967 and 1968 and titled "Thirty-One Photographs." A collection of untitled images with a defiantly bland title signaled a claim to the type of "observing and reporting" that Hoffer called being civilized. Gedney finished a handmade version of the book in October 1968; it was never published, nor does it seem like he put great effort into getting it into publishers' hands.[45]

Five images from Kentucky appeared amid the images of streets, shop windows, his grandparents, San Francisco hippies, and automobiles: the tiny Bill Cornett getting a drink at the kitchen sink, Bidge Williams stringing beans on the porch, Johnny Cornett behind the wheel of a pickup, Wilma Jean Cornett daring the viewer to blink, and the quiet embrace of Vivian and Bernice. Gedney had edited himself strictly to get down to just these five shots. He originally envisioned a forty-photograph collection that was evenly divided between San Francisco and Kentucky. But he soon took up the idea of a collection that would parallel Evans's *American Photographs*. The references to Evans's celebrated book are clear: the opening picture of a busy storefront; the "consciously impartial" take on public events such as strikes, parades, and auto shows; the scenes of individual fatigue and despair; the attention to architectural detail. Four of the Kentucky images appear together in a sequence at the beginning of the book, building from the candid shots of Bill and Bidge to the portraits of Johnny and Wilma Jean. The effect is a balance between the intimate and the standoffish. He turned these people into what the photographer Jerry Thompson called "figures of stature, figures to be looked at with wonder." Small moments loom large. The fifth Kentucky image, of mother and child, appears as the seventeenth entry in the book, the start of a three-part sequence showing people embracing. Similar sequences in the book show three "found" portrayals of Jesus, four consecutive people with their heads in their hands, four night scenes of palpable stillness, and three individuals decentered and seated in interiors that show their age. The book ends with an elderly woman looking out her front door—the figure whom Gedney would later call the "watcher of the street."[46]

We can think of "Thirty-One Photographs" as the artist measuring himself up to his idol. Unlike *American Photographs*—or Robert Frank's controversial *The Americans* from nine years earlier—Gedney's book did not try to mount a narrative about the world as he saw it. Instead, it was a narrative of his interests, his techniques, and his ability to connect formal themes. Gedney was better at creating single pictures than at editing a group to form a coherent story. This worked fine for Szarkowski, who argued that single images displayed the power of photography more successfully than series. Images could capture a series of ideas or momentary combination of forms, but Szarkowski did not think that photography constructed narratives well. Photo essays that stood out as especially evocative, he observed, relied on written captions to do much of the narrative heavy lifting. Gedney's collection of caption-free images conveyed a statement of style rather than a thesis about "American life."[47]

When the book project stalled permanently in 1968, he could not have brooded too long on it, for he had brighter prospects just ahead. In the year after his return from San Francisco, Gedney and Szarkowski conferred about a MOMA exhibition. Szarkowski pushed him toward a collection that was balanced between his two most significant projects. The resulting show, *Eastern Kentucky and San Francisco: Photographs by William Gedney*, ran from December 1968 to March 1969. This came eighteen months after what was perhaps the most influential art photography show of the 1960s, at MOMA in early 1967, as Gedney was returning from California. The *New Documents* exhibition featured the images of Arbus, Friedlander, and Winogrand, announcing a new generation of documentary photographers. Szarkowski curated *New Documents* with an eye toward the unvarnished, unembellished view of 1960s America which he believed they offered. Gedney's Guggenheim trip had aspired to such heights, but he did not have the volume of finished work to be considered for such an exhibition in early 1967. To Szarkowski, his pictures were surely of a piece with this trio of "new" documentarians. A year later, when Gedney had made dozens of prints from his San Francisco negatives, Szarkowski came calling.[48]

The MOMA show displayed forty-three of Gedney's prints, twenty-two from Kentucky and twenty-one from San Francisco. The Kentucky prints were mostly outdoor candid shots of Cornetts and Couches in everyday situations, including "Boyd on hill," and featured four of the five pictures that Gedney had chosen for "Thirty-One Photographs." Szarkowski wrote the exhibition's introductory wall label, in which he presented Gedney's work in much the same language that he had used to explain the importance of the *New Documents* artists. Following Gedney's lead, Szarkowski framed the work as studying people in a struggle with the world around them. The curator presented the subjects as "two groups of Americans who live outside the value structure that most of their countrymen consider normal." The residents of Leslie and Perry Counties were "dropped out by the choice of history," a jumble of forces beyond their control or understanding. Szarkowski then alluded to the contemporary dynamic that gave Gedney's images such force: "Both groups have been made pawns in the game of debating the state of the American experiment." These images were somehow different; they broke the hillbilly/hippie mold to expose the messiness that came with actual lived experiences. “Sunday supplements and scholarly journals,” noted Szarkowski, had turned these people into thin symbols of victimhood and protest. Gedney’s pictures restored them to “people living precariously under difficulty.” This was not more of the same, but a break from the clichés developed in the past ten years to deal with the fact of Appalachian poverty. “Rather than reminding us once more of the conventional assumptions we have been conditioned to accept,” Szarkowski argued, “these pictures reward us with real knowledge of the lives of specific people.”[49]

“Real knowledge” summarized well the value of Gedney’s pictures, at least as highbrow art circles received them in the late 1960s. *Village Voice* critic A. D. Coleman praised the exhibition for showing mountain residents’ own perspective. Coleman clarified that Gedney did not merely pass through eastern Kentucky, but rather lived with his subjects until he achieved an “objectivity through involvement.” Reviewers liked to reference such contradictions as a way of explaining the images’ sense of intimacy and distance. In the exhibition’s press release, Szarkowski told media outlets that Gedney had achieved “intimate portrayals of the people of eastern Kentucky” yet had the discipline to remain “simple, direct, and unobtrusive.” Coleman agreed, declaring that the photographer had “succeeded in getting inside” Appalachia, and if his show ultimately conveyed melancholy, it was because he caught the Couches and Cornetts in introspective moments. Gedney was able to generate visually the moods of everyday life, extraordinary only when ripped from mountain contexts and installed on the third floor of the Edward Steichen Photography Center. Coleman echoed MOMA’s copy in marveling at the “relaxed” effect of the compositions and concluding that they showed “no distortion, no attempted polemic.” Szarkowski branded Gedney a “good witness,” a practitioner who did not try to “direct our verdict concerning the quality of these lives.”[50]

The intervention that these reviewers saw in media treatments of the mountains consisted primarily of the moral judgments that Appalachian residents needed to be changed and that their lives were problems. Being a good witness meant controlling the impulse to create another “Depressed Area, U.S.A.” MOMA told its visitors that Gedney had “spent many months living among his subjects” (which was nearer to the truth for San Francisco than for Kentucky), without doing what Meyerowitz called “bruising” the situation. He did not intervene; he watched and photographed. But there was more here than mere reporting; Szarkowski stressed that the creation of the photographer—the actual work that the photographer did—was establishing the particular subject in a particular moment; he denied that there were preexisting subjects waiting for photographers to claim them. The hillbillies of Kentucky and West Virginia were created by photojournalists and their editors. Now, a new series of sui generis subjects had been created. The MOMA patron would never know the name "Boyd Couch," but the photograph of Boyd above Grassy Branch Holler hinted at the details of life in that spot. Gedney selected Leslie and Perry Counties as places worthy of investing his attention, and that counted as an undistorted view.[51]

The photography historian Gilles Mora notes that Szarkowski's influence was unprecedented in American photography circles during his first two decades at MOMA. The director's favorable opinion gained photographers access that was otherwise hard to achieve. An exhibition like Gedney's, writes Mora, "was as important for a photographer's reputation as the cover of *Life* would have been during the 1950s. It allowed photographers to be discussed at universities, to get more grants, and to increase their gallery sale prices significantly." Szarkowski viewed the role of a museum curator as providing a liaison between artists and their "potential public, including, very importantly, other artists." Gedney's Kentucky prints had attracted a modest audience before, via a one-man exhibition at the Pratt Institute in 1965, but now he attracted great, if temporary, scrutiny.[52]

In the spring semester of 1969, as his MOMA exhibition ran, Gedney began part-time teaching positions at the Pratt Institute and the Cooper Union. Meanwhile, he read the reviews pointed out by friends. In what Gedney called a "complimentary, nothing review," *Modern Photography* editor Julia Scully used a survey of his show to advise her readers on how to take more intimate photos of the people around them. Scully compared Gedney's pictures favorably to Farm Security Administration shots of "hollow-eyed" rural people and their "rag-covered" children, because Gedney's shots offered up the "slow beat" of his subjects' lives. Tedium could be beautiful. Gedney's disdain for the review's inconsequence may have stemmed from Scully's instructions to readers to be more patient and to make extended studies of the people in their lives. This could be a case in which Gedney's elitism showed. However, he must have appreciated her remark that his work did not hold the residents of eastern Kentucky up for "amusement or novelty."[53]

More valued acclaim came from the *New York Times* art critic John Canaday, who ended his review of a MOMA drawing exhibition with a nod toward the neighboring photography shows. Canaday lauded Gedney's pictures as "human documents" that were "so satisfying [a]esthetically" that they made painting seem lifeless. Gedney noted this "very handsome compliment" but admitted that it would "warm [him] for about one minute and then mean nothing." Perhaps he was buoyed, then, by Szarkowski's decision to purchase ten of the Kentucky photographs for MOMA's permanent collection. Four were candid shots of people directly in front of houses: a neighbor of the Cornetts cleaned a rifle on a porch while a young boy watched from several feet away; men sat in a yard strewn with car parts, bed posts, and an overturned chair; two teenage boys smoked on a porch; and a shirtless Cooper Blevins sat on a kitchen chair on a porch, feet up on a post, watching three dogs mill about a teenage boy. Two were posed pictures: Johnny and Wilma Jean Cornett. Four were isolated action scenes, in which the work or activity of residents filled the frame: Burchild Cornett, all ribs and dirty fingernails, carefully rolled a cigarette; two men changed a tire on a battered Chevrolet truck; a boy carried a younger boy on his back as they moved away from the camera down a dirt road; and six kids of varying age stood on a wooden bridge, fixing a bike in front of parts, cars, shacks, and trees. Two were interior shots: the fifteen-year-old Ronnie Cornett slept soundly under a hanging picture of Jesus as shepherd; and four Cornett girls stood with each other in their kitchen, framed by a fridge, a sunny window, and a small painting of the Last Supper. Finally, two were what functioned as landscapes in Gedney's Kentucky collection: an abandoned car beached in the dirt of the Blue Diamond mining camp, windows smashed, tires gone, doors and trunk open; and Boyd Couch on his hill.[54]

Photographers of talent, Szarkowski wrote in 1975, knew that "photography can help us know, rationalize, and share levels of experience that have been untouchable and uncommunicable." He liked photographers to remain hidden in their work, their choices and agendas buried under the details that made up the visual experience. Just before he left for India, Gedney wrote that he wished more photographers "would simply document instead of trying to be artists." Stating the case for simplicity, he added, "It is not easy to be unpretentious." In the five years after his Kentucky trip, Gedney had learned the language of the new documentary style. But his efforts to bring the mountain pictures to yet another public viewing show that he was not willing merely to document. He wanted to guide others in seeing what he saw; he wanted to communicate something about life in Leslie and Perry Counties which took experience to notice.[55]

As the attention of the art community shifted from Gedney in the years after the MOMA show, he revisited the mountain project, both literally and figuratively. Four months after he returned from his 1972 Kentucky trip, Gedney was ready to move forward with a reimagined book project. In the journal that he kept to annotate images, he wrote, "In looking over the Kentucky pictures of 1964 I find them generally weak and the new ones much better." A book dummy of both sets would balance what he saw as a discrepancy in quality by highlighting the time gap shown in his subjects' faces and bodies. "I only wish I had a more complete coverage in 1964," he grumbled. "There are a few of course that stand up, but I believe my most recent work is more encompassing and of all over higher quality and variety." As if willing himself back into the project, he wrote, "I am in much better control now than then (as it should be)."[56]

In the following weeks, he assembled a book dummy with both sets of photographs. He chose twenty-nine photos from 1964 for the first part of the book, including all but one of the prints that Szarkowski had kept for MOMA's collection. The images he chose to augment the original set featured six more images of people sitting or standing on porches, five more images of people leaning on or fixing cars, two more portraits of the Cornett children, and two middle-distance shots of houses in the landscape. The sequence mattered, because Gedney hoped to direct the viewer's experience through the relationship of themes in each image. The book would open with the one–two punch of "Boyd on hill" and Wilma Jean staring directly at the viewer. He planned a sequence in which the viewer moved from the porches of the Couch and Cornett houses to the cars waiting outside, with men and boys working on them, and then back to the porches, where several pictures showed parents with children in the middle of routines and idle thoughts. A series of four photos would show progressively younger sons around the house, followed by another set of three showing the jumble of junk in which his subjects lived. A few more images of people working outside preceded the last four intimate shots: the piggyback ride, sleeping Ronnie, the girls in the kitchen, and, finally, Johnny in the front seat of a truck, staring back. Gedney scribbled in the margin "ending is not right (end with kitchen)." A second draft started with the picture of the elderly woman leaning into the car window, followed by Boyd on the hill, and ended with the girls in the kitchen. A third draft put Boyd back in the first slot and ended with the elderly woman / girls in the kitchen / piggyback sequence.

Gedney second-guessed himself repeatedly during the preparation of the book. First, he thought that the size of his initial wave of prints made them look "gross." Then he criticized his images for lacking contrast. Next, he deemed his 1964 images inferior to his 1972 batch again; the former were too repetitive. Now, he was on a self-critical roll: "I looked over the 1964 contacts of Kentucky, the shooting is terrible how I ever came out with anything is a miracle. So many 'one shots,' situations left entirely unexplored. Thank God I get better the more I use the medium." Finally, he spiraled into confusion: "I go from good to bad feelings about it. somedays I doubt that the recent pictures are any good."[57]

All of this printing and editing took its toll. He complained in his notebooks about the tedious development process. Sizing and resizing the images occupied days on end, without a hint of creativity. In mid-February he sighed, "The mechanical technical side of photography takes so much time, most of it is wasteful and boring." Still, the oversupply of prints allowed him to share the wealth. When he passed through Chicago in the fall of 1975, Gedney gave Hugh Edwards, a retired photography curator of the Art Institute of Chicago, a print of Willie and his sons fixing a car. Edwards had fought the same fight as Szarkowski, defending documentary photography as an art form in the fifties and sixties. He also did for Chicago what Szarkowski did for New York: encourage local photographers to shoot in the streets and make pictures that avoided heavy-handed symbolism. Edwards wanted photographs that a dozen people could interpret in a dozen ways. Danny Lyon, whom Gedney adopted as a role model of sorts, considered Edwards a beacon in the lonely landscape of street photography. Even though Gedney connected with Edwards through the work, the project ended abruptly. The last record of his effort on the ill-fated book was a comparison to Lyon's 1971 photo book *Conversations with the Dead*, which was "aprox. the same size . . . has 196 pages and sold for $7.00 paperback."[58]

Conclusion

Years before such commercial considerations, Gedney's first audience were the Cornetts themselves. Vivian begged to see his shots several months after he left in 1964. She pressed him, writing that her family was "expecting one of [his] pictures." When he sent a large collection of photos to Big Rock that spring, she praised them heartily, offering that she did not have the money to buy enough frames. Part of Vivian's reaction was surely appreciation for the images, but part of it must have been appreciation of Gedney, too. He sent money to the Cornetts, who by that time had been without regular income for almost a year. Blue Diamond informed its workers on August 11 that they had two weeks to return to work without union representation or they would be replaced. The following day, Vivian wrote to Gedney to tell him that the family missed him, that she had just about finished her canning, and that their four-year-old son, Billy, was now calling himself "Big Bill Gedney." She treated Gedney's aid as a miracle. It also demonstrated to her that her trust in him had been warranted. "People didn't no [*sic*] what to think" about his motives, she confessed, but the money and the photographs proved that she had been right. "You opened their eyes," she wrote, "You were so kind to all of us we can never forget you."[59]

The images that Vivian adored were mostly family pictures, showing her kids around the house. Her strong reaction to the images must have been gratifying, if not unexpected, for Gedney. He knew that his audiences in Kentucky and New York were not after the same. He took to heart the photographer Charles Harbutt's explanation of the vernacular meaning of photography: "The grandmother with her box brownie knows what photography is all about. She wants to take a picture of junior because junior is going to grow up, and we're all going to get old and complicated and right now in the summer sun junior is a source of joy. Click." Gedney knew what his subjects wanted from his photography, and he discovered between 1964 and 1972 what his peers wanted. Vivian repeatedly welcomed Gedney back to Big Rock in the years after he visited, noting that their poverty would be overcome by their generosity. "As you no [*sic*] we are poor people and can't treat you like we want but we will welcome you back any time," she promised. It was eight years before he took Vivian up on the offer.[60]

Boyd Couch also reached out, but he was not interested in pictures. Couch wrote in January 1965 to see whether Gedney might spread word about the miners' plight to his friends in New York. Unemployment payments had come to an end, and the UMW local needed sympathetic benefactors to sustain the picketing. The picketers engaged in violent clashes with strikebreakers in the winter and spring of 1965, culminating in the shooting death of thirty-nine-year-old Ernest Creech as he left the mine on March 3. The union picketers were rounded up and arrested, including Boyd Couch, who had recently moved out of the Leatherwood miners' housing to Delphia, 5 miles away. Gedney seemingly never responded to Couch's request.[61]

The scholar Katherine Ledford asks whether there is a difference between a "stranger with a camera" taking one's picture and a "second cousin" doing the same. Does familiarity matter in an act of representation? Ledford's question concerns reception—what might people *think* about photographs taken in the mountains of Appalachia?—and frames it in terms of the subject's relation to the photographer. That is one way of considering the insider/outsider relationship. Gedney might have had technical skill that Willie or Vivian could not have brought to a camera, but would family pictures taken by the Cornetts have *meant* something different than Gedney's images? Through a lens focused on the wide range of positions of which individuals could be inside and outside, the issue of how others received Gedney's pictures becomes a way to think about the use of photography. Or, as Szarkowski put it, "What are photographers for and how are they to be used?" Vivian liked Gedney for what his work said about her and her family, as well as for what he could do for them. Gedney's photography peers did the same. The declared authenticity of the Kentucky photos attested to the power of street photography to master the ephemeral and the power of art to master what the popular media could not. He evaded what he called the "torture chamber" of media-induced conformity to message and method.[62]

He wrote back to the Cornetts upon returning from India, though he never received a reply. The following year Gedney sold his first image from the 1964 trip for a book about photography and children. The photo was of the Cornett girls in their kitchen. He sent $35 to Big Rock in June 1972, reminding them that he had promised to split any money he earned from the pictures. He asked how they were and apologized for falling out of contact. "Maybe I am a stranger to you now," he wrote; "I hope you still think of me as a friend." Vivian replied with gratitude and an apology, saying that her "nurves" were so bad that it was difficult to write letters. The family had expanded, with Vivian having her last child three years before and four grandchildren keeping three of the Cornett siblings busy. Another grandchild "got killed in a wreck" two weeks earlier. And Willie was advancing through the screening process that would qualify him for black lung benefits under the Federal Coal Mine Health and Safety Act of 1969. "It is hard living on food stamps," Vivian confided, and she thanked Gedney for the money. "Bill why I didn't write and tell you to come on down we didn't have anything hardly to eat we don't have much yet but come on down and we will fair alike we are poor as you no but we will treat you good as we can we would love to see you we think of you often." Two months later, Gedney was back in Big Rock.[63]

Shortly after he returned to New York from the second trip, Gedney received another letter from Vivian. She glowed in the aftermath of his visit: "Bill, why are we so attached to you? You feel so close somehow, like you are close kin to us. Sorry you had to fare so bad while you were here. It is really hard with the family I have, I just can't treat people like I want to.

But I really love to have you with us, it is a real pleasure. We enjoyed every moment you spent with us." The same October 1972 letter in which Vivian declared her family's love disclosed that Ronnie had been stabbed by Rufus Baker and needed to stay in a hospital "for some time." "Rufus cut his guts out," she wrote, "also someone burned Ronnie's house and everything he had that wasn't much as you no [*sic*] but it hurts to lose what you have hope he will settle down after this." In March 1973 Gedney sent Vivian Cornett eighteen prints that he described (to himself) as "8½ × 11 rejects." These were shots that he had made seven months before. They included candid shots of children giving each other piggyback rides and playing in the front seat of a truck, portraits of smiling children, images of Willie snapping beans on the porch, Junior reclining on the hood of a car, and a group shot of the family posed on the front porch.[64]

As he abandoned his planned Kentucky book in 1973, Gedney surveyed the field of photography in the age of ubiquitous television and glossy photo books. Photographers could not escape the past, because they saw the styles of past masters everywhere they went. "Too accessible a history becomes a burden," he wrote, "like always having your father looking over your shoulder." The mass media's image makers had such power to deal with the "content" of daily life that they forced practitioners like himself to "deal with form and style" exclusively. Two years earlier, when Diane Arbus had committed suicide in August 1971, Gedney remembered that she had called the figures from his Kentucky photos "his people." If he made such a claim to the Couches and Cornetts, it was less about the content of their lives—despite his references to their hardships and idiosyncrasies—than about the moments that he created with them out of the raw material of the everyday. He made strides with the products of his mountain visit because he convinced his colleagues that his journey to the margins had "rendered technical" the contours of life in Appalachia. Images devoted to form rather than content, to routines rather than cultures, suggested an unemotional sensitivity that seemed like a particularly apt way to encounter Appalachia in a post-JFK world.[65]

Conclusion

Marshall Everett Vaughn, the founding editor of *Mountain Life and Work*, told readers in 1925 that the goal of his magazine was to combat the "unauthoritative characterizations by professional writers and unscientific reformers" which had plagued southern Appalachia. Berea College fostered the magazine as an initiative to develop new, more flattering publicity about the people and the land. The "best thinkers on mountain life" would reach both residents and nonresidents with a message of the mountains' potential. At the time that Vaughn declared this intention, J. Horace McFarland had retired from two decades of promoting the mountains that sat 500 miles northeast of Berea's campus. In 1925, Benton MacKaye, 250 miles further to the east, was halfway between his announcement of the Appalachian Trail concept and the publication of *The New Exploration*. The Smoky Mountains hikers were in their clubs' honeymoon phase, still figuring out what was possible on the peaks between Knoxville and Asheville, 150 miles south of Vaughn's office. Marion Pearsall had just turned two years old in April 1925; William Gedney would not be born for another seven years. Separated by distance and time, the experts of *Proving Ground* did not set out to publicize Appalachia. But like Vaughn, they all thought about what it meant to be in the mountains. They thought about themselves, and they thought about how their experience of the mountains could serve fellow specialists.[1]

Vaughn wrote of a simplicity to Appalachia that could "help to sober the dizziness of our national life." In certain hands, the stereotype of simplicity could become a patronizing excuse for manipulation, a way of explaining why particular types of economic development had to happen for the good of residents. A retreat to simplicity sold the complexity of the mountains short. But there were other ways to treat dizziness. Each of the experts discussed here found a rich intricacy in the mountains. They did not try to encapsulate their experience in aphorisms or suggest that what they did on high was easy to replicate. Others could achieve what they achieved, but it would take a commitment to their craft. If they saw the Appalachian Mountains as a remedy for what was ailing the United States in the twentieth century, it was through visitors dealing with invigorating beauty, complex natural processes, and thought-provoking social change. They staked their professionalized claims on the notion that what they were doing was difficult. They reached for the type of "second level of observation" that Appalachian Studies scholars pursue in order to avoid one-dimensional studies. None of them thought that simplicity had anything to do with it.[2]

In the annals of professions as divergent as regional planning and art photography, the instinct to reach out-of-the-way spaces to make one's mark persisted in the twentieth century. McFarland's brand of scenic preservation was by no means new in the 1910s, but his merging of preservation, automobility, and photography made Eagles Mere just far and wild enough to present himself as trailblazing. MacKaye seized upon the very concept of topographical remoteness at a time when regional planning needed captivating ideas and defining images. Concerted, regimented hiking in the Smoky Mountains became a phenomenon only through the actions of the clubs established in the 1920s. Medical anthropology did not exist as an academic field when Pearsall stayed in Leslie County in 1959; her work, three hours out of Lexington, allowed her to become an authority in a discipline that was founded around such experiments. Finally, one could certainly be a successful street photographer without ever leaving New York City. But Gedney earned professional attention by applying the tenets of street photography to the unexpected, distant locales of Leslie and Perry Counties. Each of these practitioners got something out of their projects, even as they offered themselves as experts for colleagues or potential followers. Their professional networks were animated by a belief in inventing new ways of experiencing the world through mobility. Road-building programs and the spread of private vehicles broke down the remoteness of the mountains over time, but the sites visited herein remained firmly within a "special frontier . . . a frontier-inside" that was imaginary (based on the narratives of occupational cultures) and tangible (experienced through visitors' bodies as they walked, climbed, drove, ate, photographed, etc.).[3]

Each of these figures argued that there were no short cuts; time spent in the mountains was the only way to get better at their practices. The time that they actually devoted to "fieldwork" was usually quite short. Pearsall's half year in Hyden was the longest stretch by far, though the would-be experts accrued time in the mountains through successive trips. Time was valuable, of course, and the privilege of spending it in the mountains united these travelers. McFarland's affluence was exceptional among them, but the rest were at least materially comfortable. Both MacKaye and Gedney suffered bouts of deprivation in their careers, but they had professional and private networks of people to whom they could always turn for a temporary gig or a couch on which to sleep. Mobility was a luxury, even when portrayed as trial and torment.

Historian Penny Messinger notes that the concept of "social location" can help scholars push through the dilemmas of insiders versus outsiders and Appalachia-as-fantasy versus Appalachia-as-reality. Social location relies on a sense of place to inflect other facets of identity, with that sense deployed variably in different social contexts and at different times. Social location involves such variables as class, gender, race, and age channeled through embedded experience on the ground. And senses of place are likewise affected by individuals' positioning in multiple social relationships. Visitors to the mountains had social locations, too, and they included Appalachian locales experienced in particular ways at specific moments in personal and occupational histories. Claiming expertise via Appalachia (a lay knowledge of the mountains) was a "territorially embedded and culturally embodied experience." Their will to improve in the mountains relied on a sense that the mountains were not finished, that more could be said and done within them. Thus, there was a chance for them to prove themselves.[4]

In a recent review of the shifting frameworks of Appalachian Studies scholars, William Schumann summarizes the view from the field in the 1970s as a "picture of complex, systemic class-based control operating across a range of social, economic, and political fields and at different scales of power." Although the discipline has moved in many directions since that time, the specter of the controlling intervenor retains a powerful presence in research about the interactions of residents and nonresidents. Appalachian Studies writers have shown that local people managed to control aspects of their lives, even as other parts of their lives were determined by absentee capitalists and government agencies. Intervenors often represent the capitalists and agencies; in scholarly analyses, they have tended to serve as the means by which places were compromised or destroyed.[5]

Each of my episodes was an effort at place making, a process through which visitors claimed specific geographic spaces as occupational zones for burgeoning fields. The cultures that they attached to their chosen destinations were the cultures of professions (or professionalized leisure) as they extended into remote locales. The remoteness of the locales, when measured in distance from the core and by the logistical difficulty surmounted during the trip, mattered. They were "doing place," then, but the places they practiced were not Appalachian places as such; they were the grounds of occupational development. Experts improve their chances of being believed by mastering the art of reframing situations as problems with specific solutions. This was a convenience for the practitioners, a way of getting out of the mountains exactly what they wanted to get out of the mountains. And some would argue that a convenience for one is exploitation of another. If they were interested in mountain residents at all, they were interested in them as case studies—projects that could move their disciplines forward. They exhibited varying levels of detachment to Appalachian residents and were explicit about their goals and outlooks.[6]

Of these people, Pearsall was most attentive to the cultures of the mountains, though she understood them as local frames that were not generalizable. She adopted the approach to culture which recent writers have characterized as "belief and behavior sets tied to specific places." By avoiding overblown claims to typicality, Pearsall carved out a niche for herself in the literature of social science methodology. She earned a spot for herself in the reflexive study of "mountain work." Her project thus serves as a good model for understanding the efforts of this book's agents. They focused on themselves as guests in the mountains, representatives of pursuits that valued initiative and the ability to deal with complexity. The Appalachian Mountains provided the novelty and uncertainty against which they had to perform. The performances were alternately aesthetic, intellectual, and kinesthetic, but they all involved the sensuous experience of Appalachian landscapes.[7]

All of these figures presented themselves as experts in mountain practices. McFarland knew where and how to look in order to appreciate the upland outdoors in the "right" way. MacKaye knew how to read the mountain landscape to produce new stories to guide lowlanders' sense of self. Leaders of the Smoky Mountains Hiking Club and the Carolina Mountain Club knew how to elevate mountain mobility to a state of stimulating flow. Pearsall knew how to stake a social scientist's claim to a discrete section of the mountains and methodically turn it into a unit of analysis. And Gedney knew how to hover on the fringes of mountain residents long enough and quietly enough to picture their lives as aesthetic moments. The expertise on display was different in each case, but their common ground was rendering the Appalachian Mountains as a technical terrain. The devices by which they rendered Appalachia technical were photographs, lists of instructions, maps, contact sheets, reports, field notes, diagrams, sketches, and inventories of supplies, trails, and rolls of film. These were the tools of their trades, so to speak, and creating them confirmed the mountain visitors as members of the credentialed cohorts that they prized.

By focusing on the material practices of these five projects, this book has studied the building of relationships with the land. The details of how these travelers produced their stories of the mountains were constitutive; they were methods of communicating ideas about experiences in particular places and times to audiences that were not, and might never be, in those spots. And as origin stories about types of expertise, they showed how practitioners forged something new by working out problems that they had created for themselves. Their framing of those problems both reproduced and destabilized such worn binaries as city/wilderness and progress/stagnation. The difficulty that the agents needed to relay to others echoed views of Appalachia as a land apart. Yet their faith in the portability of their instructions—the paths toward achievement that they laid out for colleagues—insisted that being in the mountains could become customary for those who trained enough. They showed that their Appalachian sites were spaces of practice that could enrich others' lives. The fact that those "others" were people like themselves, working in the woods and then returning to urban routines, was an inherent part of their claims to expertise.[8]

NOTES

Introduction

1. Marion Pearsall field notes, June–July 1959, box 10, Marion Pearsall Collection.

2. Whisnant, *All That Is Native and Fine*, 119; Ronald L. Lewis, "Beyond Isolation and Homogeneity: Diversity and the History of Appalachia," in *Confronting Appalachian Stereotypes*, ed. Billings, Norman, and Ledford, 22; Dunn, *Cades Cove*, 144; Jordan-Bychkov, *Upland South*, 3; Batteau, *Invention of Appalachia*, 100; Shapiro, *Appalachia on Our Mind*, 217. On the concept of a "socially-maintained reputation of a place or region," see Shields, *Places on the Margin*, 13–14.

3. Russell, *Looking North*, 57.

4. Huntley, *Making of Yosemite*, 9.

5. Berleant, *Aesthetics of Environment*, 5, 155; David Crouch, "Surrounded by Place: Embodied Encounters," in *Tourism*, ed. Coleman and Crang, 212; Edward Casey, "How to Get from Space to Place in a Fairly Short Stretch of Time," in *Senses of Place*, ed. Feld and Basso, 18, 43; Entrikin, *Betweenness of Place*, 57.

6. Wiebe, *Search for Order*, 113, 121, 129.

7. D. Mitchell, "Lure of the Local," 271. On the stigma of the tenderfoot, see Turner, "From Woodcraft to 'Leave No Trace,'" 465.

8. "Everybody gazes at everybody," writes Darya Maoz. Maoz argues that locals adopt savvy strategies to defend themselves against—or profit from—the tourist's gaze. Ultimately, she notes, the power that locals gain from observing travelers and creating appropriate displays for their consumption is a partial, hollow benefit. See Maoz, "Mutual Gaze," 225, 234–35.

9. These ways of viewing the outdoors are from Meinig, *Interpretation of Ordinary Landscapes*, 33–48.

10. Russell, *Looking North*, 48; Marx, *Machine in the Garden*, 43; Batteau, *Invention of Appalachia*, 16. The sixties-era stereotyping was nothing new, as historian David Hsiung has shown by examining eastern Tennessee in the late nineteenth century. High-mountain residents who foresaw little direct benefit from the encroachment of railroads contested transportation projects. Their immediate neighbors, whose stores, farms, and communities lay along prospective routes, championed the railroads while simultaneously painting the "branchwater mountaineers" as isolated and ignorant for their opposition. See Hsiung, *Two Worlds*, 159. On the invention of Appalachia as a conceptual

region, see Shapiro, *Appalachia on Our Mind*; Whisnant, *Modernizing the Mountaineer*; and Whisnant, *All That Is Native and Fine*.

11. Powell, *Critical Regionalism*, 4, 7, 21; Stradling, *Making Mountains*, 122.

12. R. D. Mitchell, *Appalachian Frontiers*, 2; Whisnant, *All That Is Native and Fine*, 260; R. L. Lewis, *Transforming the Appalachian Countryside*, 19.

13. Whisnant, *All That Is Native and Fine*, 265; Feintuch, "End of an Era," 105; Satterwhite, "Imagining Home, Nation, World," 30; Pierce, *Great Smokies*, 196; Judy Bonds quoted in House and Howard, *Something's Rising*, 144.

14. Cronon, "Trouble with Wilderness," 86; Benjamin Johnson, "Wilderness Parks and Their Discontents," in *American Wilderness*, ed. M. Lewis, 113–30; Casey, "How to Get from Space," in *Senses of Place*, ed. Feld and Basso, 35.

15. Williams, "Appalachian History," 175; Li, *Will to Improve*, 7, 126; Donald Edward Davis and Chris Baker, "Fixing Appalachia: A Century of Community Development in a 'Depressed' Area," in *Studying Appalachian Studies*, ed. Berry, Obermiller, and Scott, 89–101.

16. Nye, *American Technological Sublime*, 23. On the instruction of visitors in rural spaces, see Parker, "Country Code," 1–16.

17. Crouch, "Surrounded by Place," in *Tourism*, ed. Coleman and Crang, 217; Vivian Cornett to William Gedney, 25 October 1972, Correspondence, William Gedney Collection. See Herbert Reid's helpful consideration of insider/outsider traps in Appalachian Studies: Reid, "Appalachia and the 'Sacrament of Co-Existence.'"

18. White, "From Wilderness to Hybrid Landscapes," 558; Shaunna L. Scott, Phillip J. Obermiller, and Chad Berry, "Making Appalachia: Interdisciplinary Fields and Appalachian Studies," in *Studying Appalachian Studies*, ed. Berry, Obermiller, and Scott, 30; Stokowski, "Languages of Place," 369; Huntley, *Making of Yosemite*, 3–4; Stroud, *Nature Next Door*, 23; Anna Rachel Terman, "Intersections of Appalachian Identity," in *Appalachia Revisited*, ed. Schumann and Fletcher, 82–83.

19. Among the works that focus closely on nature–culture hybrids are Hou, *City Natural*; Enright, *Maximum of Wilderness*; Armitage, *Nature Study Movement*; and Van Slyck, *Manufactured Wilderness*.

20. Nash, *Wilderness and the American Mind*, 51, 57. One scholar who has considered occupational connections to commentary on Appalachia is Jeffrey Guy Johnson (see "Feud, Society, Family," esp. 241–48). Johnson demonstrates that some turn-of-the century legal observers saw the argumentation of the courtroom as a form of combat that echoed the Appalachian feud narratives of the late nineteenth century. The American "cult of professionalism," noted Johnson, was new enough that its creators sought such analogies as useful frameworks for occupational identity.

21. Crouch, "Surrounded by Place," in *Tourism*, ed. Coleman and Crang, 208–11.

22. Kathleen Stewart, "An Occupied Place," in *Senses of Place*, ed. Feld and Basso, 143; Cosgrove and della Dora, *High Places*, 4.

23. Taylor, *Dream of England*, 15; Armitage, *Nature Study Movement*, 37.

Chapter 1 · Expert Vision: J. Horace McFarland in the Woods

1. J. Horace McFarland (hereafter JHM), "Bringing the Folks to the Forest," 1 February 1929, New England Forestry Congress, JHM Papers; JHM, "Eagles Mere This Summer," ca. 1910, JHM Papers.

2. *Harrisburg Telegraph*, 14 May 1915, 16.

3. Lamme and Russell, "Removing the Spin," 338–41. See also Lamme, "Public Sentiment Building Society," 123–32; and Cutlip, *Public Relations History*.

4. JHM, *Floral Designs*; "J. Horace McFarland," 436; Morrison, *J. Horace McFarland*, 29, 57, 98; Knight, "Environmentalism of Edward Bok," 159; *National Nurseryman*, 184; "Harvard's Course in Printing," 913.

5. Wilson, "J. Horace McFarland," 315. Wilson notes that McFarland delivered his motivational "Crusade against Ugliness" speech at least 250 times in the decade after the Harrisburg campaign, typically charging $50 or $75 per appearance. On the Harrisburg reform community, see Rimby, *Mira Lloyd Dock*.

6. Hays, *Conservation and the Gospel of Efficiency*, 194–95; Peterson, "Birth of Organized City Planning," 127; Armitage, *Nature Study Movement*, 10; JHM, "Why Billboard Advertising," 20; JHM, "Bringing the Folks to the Forest"; JHM, "Pennsylvania's Scenic Supremacy," radio address on WBAK Harrisburg, 26 November 1928, JHM Papers.

7. Young, "Social Reform through Parks," 461; JHM, "Shall We Have Ugly Conservation?," 596; Knight, "Environmentalism of Edward Bok," 159; JHM, "The Wrong Education" (March 1906 draft of "Education in Ugliness"), JHM Papers; JHM, "Twenty Years of Scenery-Saving in America," 306–7; JHM, "Shall We Make a Coal-Pile of Niagara?," 19; JHM, "Are National Parks Worth While?," 10; JHM, abstract of an illustrated address on "The Crusade against Ugliness," JHM Papers; JHM, "Awakening of a City," 1931. McFarland appealed to politicians by emphasizing the things that would keep them in office. He promoted a national park bureau to President Taft in 1911 by arguing that urban park systems improved "community values" and raised tax revenues along their borders. Taft could win himself influential friends wherever he promoted parks. In 1918 McFarland buttonholed Felix Frankfurter, the head of the War Labor Policies Board, on behalf of the Commission on Living Conditions of War Workers. If Frankfurter wanted to improve workers' "stability" and utility, McFarland urged, then the "amenities of life need[ed] to be considered in connection with the efficiencies of life." Immerse war workers in pleasant workplaces and housing, and they would meet any quotas asked of them. See JHM, "Are National Parks Worth While?," 12; Felix Frankfurter to JHM, 10 November 1918, JHM Papers; and JHM to Felix Frankfurter, 12 October 1918, JHM Papers.

8. Rimby, *Mira Lloyd Dock*, 24, 42; Dear, "Mysteries of State," 207; JHM, "Statement in Regard," 259.

9. JHM, "Why Billboard Advertising," 37; JHM, "Militant Citizenship," baccalaureate address to Christian Association of Lebanon Valley College, Annville, PA, 9 June 1907, JHM Papers; JHM, "Ignorance of 'Good' Citizens," 273; "At the Men's Convention in Harrisburg, Pa.," 8; Stehr and Grundmann, *Experts*, 35; JHM, synopsis of address, "See Pennsylvania First," Engineer's Society of Central Pennsylvania, 16 April 1915, JHM Papers; JHM, "Shall We Make a Desert of America?," April 1908, JHM Papers; Wilson, "Harrisburg's Successful City Beautiful Movement," 216–17; Morrison, *J. Horace McFarland*, 70–71.

10. JHM to Daniel Carter Beard, 13 November 1915, JHM Papers; Watrous, "Civic Art and Country Life," 191. See the related description of McFarland as a "lay advocate" in Chambliss, "Perfecting Space," 487.

11. Dear, "Mysteries of State," 209, 222. For a discussion of expertise as it emerged in the realms of horticulture and landscape architecture, see Hou, *City Natural*, 55–57.

12. JHM, "Shall We Have Ugly Conservation?," 594; Benjamin Johnson, "Wilderness Parks and Their Discontents," in *American Wilderness*, ed. M. Lewis, 113–14; Hays, *Conservation and the Gospel of Efficiency*, 145; *St. Louis Republic*, 30 April 1904, 8. Cutting down trees and creating scenery were parts of a whole. McFarland claimed that forests had been "provided by the Creator for the resting of tired brains and the healing of ruffled spirits, as well as for utility." National forests should be cultivated with calculated efficiency, producing marketable timber quickly by following "purely economic" principles. The economics of national parks, on the other hand, related to their effects on the people who visited them; park guests were so invigorated by natural wonders that they returned to their lives with renewed drive and productivity. See JHM, *Getting Acquainted*, v; JHM, "See Pennsylvania First"; and JHM, "Shall We Make a Desert of America?" For a discussion of spiritual traditions in American environmental thought, see Dunlap, *Faith in Nature*.

13. JHM, abstract of an illustrated address on "The Crusade against Ugliness"; JHM, "Wrong Education"; JHM, "Pennsylvania's Scenic Supremacy."

14. JHM to Allen Chamberlain, 23 December 1914, JHM Papers; JHM, "Bringing the Folks to the Forest"; JHM, "Shall Our Country Roads be Beautiful or Ugly?," 369; JHM, "What About the Country Highway?," 221, 225.

15. "Extracts relating to the zoning address to the Harrisburg Rotary Club, Monday, 26 January (1931)," JHM Papers; JHM to Mira Lloyd Dock, 28 May 1903, JHM Papers; Morrison, *J. Horace McFarland*, 73; "Dr. J. Horace McFarland," 559; JHM to P. L. Buttrick, 19 July 1920, JHM Papers; Buttrick, "Tree-Planting Plans," 11–12.

16. JHM, *Eagles Mere*, 17; JHM, "A New Eagles Mere: Some Letters that Might be Written," 24 March 1910, JHM Papers; JHM, "Pennsylvania's Scenic Supremacy"; JHM, "How Can the Association Become a More Vital Factor in the Physical Life of the Community?," 21 February 1908 speech to state YMCA at Connellsville, JHM Papers; JHM, "Summer Lake Carnival," 9; JHM, "Highest Development of the Summer Camp," 20.

17. JHM, *Eagles Mere*, 18; JHM, "Summer Lake Carnival," 9; JHM, "Eagles Mere This Summer"; JHM, "A New Eagles Mere"; JHM, "See Pennsylvania First."

18. On the social authority and "mediator status" derived from stories of travel, see Sørensen and Stepputat, "Narrations of Authority and Mobility," 313–42.

19. JHM to John Gries, 5 November 1924, JHM Papers; Morrison, *J. Horace McFarland*, 241; JHM, "Pennsylvania's Scenic Supremacy"; Mom, "Civilized Adventure," 163. John Urry argues that car culture "*coerces* people into an intense *flexibility*." The division of lives into work, domestic, consumer, and recreation spaces, all separated by great distances, forced motorists to manage time and space in new ways. Cars made possible a desire for individual flexibility in travel which soon became fraught with social pressures. Examples of those pressures were the new aesthetic rules that automobility fostered. Mimi Sheller examines the lived experience of car use, particularly the ways in which driving elicits "aesthetic, emotional, and sensory responses." She posits that when it comes to cars, cultural rules about how one should sense the world in motion also become fundamental rules about feeling. See John Urry, "The 'System' of Automobility," in *Automobilities*, ed. Featherstone, Thrift, and Urry, 28; Mimi Sheller, "Automotive Emotions," in *Automobilities*, ed. Featherstone, Thrift, and Urry, 222, 226.

20. *Motoring with a Kodak*, 3, 6. On the sensations of speed and danger in recreation, see Bell and Lyall, *Accelerated Sublime*.

21. JHM, "Bringing the Folks to the Forest."

22. JHM, "Pennsylvania's Scenic Supremacy"; Mom, "Civilized Adventure," 166.

23. Jenkins, *Images and Enterprise*, 112; "Hand Cameras," 100; Stilgoe, *Landscape and Images*, 299; JHM, "Summer Lake Carnival," 37; "He Helped to Save Niagara Falls," 865.

24. Urry, *Tourist Gaze*, 128; Blair, *William Henry Jackson's*, 57, 73, 75.

25. JHM to Mira L. Dock, 24 September 1898, JHM Papers; JHM, "See Pennsylvania First"; Orvell, *Real Thing*, 78; Cohen, *Pathless Way*, 241. Eastman Kodak counseled, "You cannot get up any great enthusiasm over the little pictures of mountain scenery—unless you have been in the same place; but let us see you go over your own collection after you have made it and see if you haven't a rhapsody or so to get off about the beauty of some sunset or the magnificence of some view." See *Motoring with a Kodak*, 9.

26. Claudy, "Lantern at Home," 64; "Abuse of the Optical Lantern," 26.

27. "The J. Horace McFarland Co.," *National Nurseryman* 16 (July 1908): 243; JHM, "Statement in Regard to the Lecture on Color Photography," *Journal of the Engineers Society of Pennsylvania* 1 (July 1909): 100; *Cyclopaedia of American Biography*, vol. 17 (New York, 1920), 218; JHM, "Horticultural Photography," 2600; JHM, "Romance of Color Photography," 298, 337.

28. JHM, *Photographing Flowers and Trees*, 40, 42; "Outdoor Backgrounds," 145.

29. JHM, *Photographing Flowers and Trees*, 51–52. On masculinism in turn-of-the-century geography, see Morin, *Frontiers of Femininity*; Rose, *Feminism and Geography*; and Kolodny, *Lay of the Land*.

30. Schrepfer, *Nature's Altars*, 42; Beardsley, "Automobile and the Beginner," 264–65.

31. *Williamsport Sun-Gazette*, 13 September 1937. The vista was officially renamed "Wright's View" in 1929, after the state's highway commissioner Paul Wright ordered the road paved so that more tourists could access the view. See *Altoona Mirror*, 10 July 1929.

32. JHM, "Highest Development of the Summer Camp," 20.

33. JHM, "Eagles Mere This Summer"; JHM, "A New Eagles Mere"; JHM, "Summer Lake Carnival," 9; JHM, "Hunting in the May Woods," 190.

34. "Print Criticism," 37.

35. "Lantern Slides," 100.

36. Torrey, Place, and Dickinson, *New York Walk Book*, xiii; Dann, *Across the Great Border Fault*, 30; JHM, *Photographing Flowers and Trees*, 4.

37. *Photo-Era* 26 (February 1911): 97; Lee, "Snapshots," 254. Hunting metaphors for outdoor photography, Kevin Armitage argues, were part of a campaign to defend the technology from claims that it alienated the individual from nature and rendered the outdoorsman passive. See Armitage, *Nature Study Movement*, 161.

38. JHM, "Romance of Color Photography," 337.

39. Ibid., 336.

40. JHM, "A New Eagles Mere"; Orvell, *Real Thing*, 99–100. For examples of such photographic advice, see Beardsley, "Automobile and the Beginner," 264–65; and Field, "Appalachian Mountain Club," 190.

41. JHM, "Highest Development of the Summer Camp," 18.

42. Sutter, *Driven Wild*, 83; JHM, "Highest Development of the Summer Camp," 18, 20; JHM; "Pennsylvania's Scenic Supremacy"; JHM, "What About the Country Highway?" 225. Ronald Foresta makes a similar point about the development of the Appalachian Trail in Foresta, "Transformation of the Appalachian Trail," 78.

43. JHM, "Highest Development of the Summer Camp," 18; Morrison, *J. Horace McFarland*, 29; JHM, "A New Eagles Mere"; JHM, "Summer Lake Carnival," 37; JHM, "Shall We Have Ugly Conservation?"; Johnson, "Wilderness Parks," in *American Wilderness*, ed. M. Lewis, 122. Johnson notes that conservationists of the era tended to see inhabited woodland as inefficiently managed and thus tried to minimize reminders of human habitation. See the related discussion about New England tourism in Judd, *Common Lands, Common People*, 226–28.

44. Wilson, "J. Horace McFarland," 326–27.

45. Cosgrove and della Dora, *High Places*, 223; Cronon, "Trouble with Wilderness," 80, 89.

Chapter 2 • Terrestrial and Human: Benton MacKaye's Mountain Work

1. Benton MacKaye (hereafter BM) to Hazel MacKaye, 21 September 1933, MacKaye Family Papers; Anderson, *Benton MacKaye*, 243, 246; "Lookout Mountain Reached by Incline Railway," *Electric Traction* 18 (April 1922): 319.

2. BM to Hazel MacKaye, 1 July 1934, MacKaye Family Papers.

3. The philosopher Yuriko Saito explains this difference as one between people who seek entertainment in nature (even entertainment of an educational variety) and those "willing to acknowledge and appreciate the diverse ways in which nature speaks." The former group perceives the outdoors as a visual resource, whereas the latter pushes past the "sensuous surface" to examine "origins, functions, and mechanisms." See Saito, "Aesthetics of Unscenic Nature," 103–4.

4. Ness, "Path Taken," 42; Anderson, *Benton MacKaye*, 48.

5. McFarland and MacKaye were not colleagues in any literal sense, but their work in publicizing the Appalachian Mountains drove similar developments. McFarland saw MacKaye as one of several people who could carry the torch for the conservation movement into the mid-twentieth century. After he had stepped down from leading the ACA in 1924, McFarland wrote admiringly of MacKaye as a lobbying voice. Moreover, he appreciated the AT project, giving it "three cheers" in a letter to Harlan Kelsey of the Appalachian Mountain Club. If MacKaye knew of McFarland's work, it would have been through the ACA's efforts to protect wayside scenery from commercial development. They attended at least one conference together, in Hartford in 1929, though it is unclear whether they were formally introduced. See JHM to Harlan Kelsey, 26 December 1924, JHM Papers; BM to Clarence Stein, 22 January 1922, MacKaye Family Papers.

6. Anderson, *Benton MacKaye*, 97, 99; BM, "Appalachian Trail: A Project," 325–30; BM, "Progress toward the Appalachian Trail," 244–45; BM, "Appalachian Power," 619; BM to Guy Frizzell, 23 June 1934, MacKaye Family Papers.

7. "Battle of Doodletown" manuscript, MacKaye Family Papers; BM, *New Exploration*, 200; Luccarelli, "Benton MacKaye's Appalachian Trail," 208–9.

8. Paul Fink to BM, 7 February 1925, MacKaye Family Papers.

9. BM to Clarence Stein, 22 January 1922, 10 February 1922, 26 February 1922, 5 March 1922, 18 November 1924; BM to Lewis Mumford, 9 March 1927; Paul M. Fink to BM, 25 November 1924, all MacKaye Family Papers. On the bustling rise of hiking clubs in the early twentieth century, see Chamberlin, *On the Trail*, chap. 3.

10. BM to Clarence Stein, 6 January 1923, MacKaye Family Papers.

11. BM, "Progress toward the Appalachian Trail," 246–47; BM to Allen Chamberlain, 21 September 1925, MacKaye Family Papers; BM to Lewis Mumford, 3 December 1926, MacKaye Family Papers.

12. Simo, *Forest and Garden*, 86; BM to Clarence Stein, 19 May 1930, MacKaye Family Papers; ATC, Report of the Fifth Annual Meeting, June 1931, Appalachian Trail clipping file, Paul M. Fink Papers; Clarence Stein, answers to Tennessee Valley Authority questionnaire, 7 October 1933, MacKaye Family Papers; Foresta, "Transformation of the Appalachian Trail," 76.

13. BM to Clarence Stein, 8 March 1925, MacKaye Family Papers.

14. BM, "End or Peak of Civilization?," 444; BM, "Boys' Project for a Balanced Civilization"; BM to Norman Thomas, 26 March 1925, MacKaye Family Papers.

15. BM to Clarence Stein, 15 November 1922, 19 October 1921, MacKaye Family Papers; BM, "To Keep Malignant Growth off Our Highways," 21 February 1928, box 160, folder 21, MacKaye Family Papers; Minteer, *Landscape of Reform*, 98; BM, "Appalachian Power," 618–19; BM to Harlan Kelsey, 31 March 1925, MacKaye Family Papers; BM to H. G. Merriam, 7 April 1932, MacKaye Family Papers.

16. Easterling, *Organizational Space*, 28; BM to Charles S. Bird Jr., 9 January 1928, MacKaye Family Papers.

17. Daniel Beard to JHM, 16 November 1915, JHM Papers; BM, *New Exploration*, 22, 26; Minteer, *Landscape of Reform*, 99.

18. Lewis Mumford to BM, 12 March 1927; Clarence Stein, answers to Tennessee Valley Authority questionnaire, 7 October 1933; BM to Charles S. Bird Jr., 9 January 1928; BM to H. G. Merriam, 7 April 1932, all MacKaye Family Papers.

19. BM, "Appalachian Power," 618–19; BM, "Wilderness Ways," 248; BM to Clarence Stein, 15 November 1922, MacKaye Family Papers; Laurence Todd to BM, 14 November 1928, MacKaye Family Papers.

20. Sutter, *Driven Wild*, 31, 70; BM, "Wilderness Ways," 243.

21. BM to Clarence Stein, 15 November 1922; BM to Harlan Kelsey, 31 March 1925; Lewis Mumford to BM, 12 March 1927, all MacKaye Family Papers; Haney, *When Modern Was Green*, 122, 124, 176; BM to Harry Dow, 22 December 1928, MacKaye Family Papers; BM, *New Exploration*, 126–27.

22. BM to Lewis Mumford, 4 April 1927; Lewis Mumford to BM, 26 July 1927, both MacKaye Family Papers; Anderson, "Benton MacKaye's 1904 White Mountains Hike," 157, 166.

23. James MacKaye to BM, 20 October 1928; Raphael Zon to BM, 20 October 1928, both MacKaye Family Papers.

24. BM to Mortimer Proctor, 20 January 1929, MacKaye Family Papers.

25. BM to Clarence Stein, 15 November 1922, MacKaye Family Papers; *Fitchburg Sentinel*, 5 October 1927, 7.

26. Lewis Mumford to BM, 18 December 1924, MacKaye Family Papers; Simo, *Forest and Garden*, 83; BM to Percy MacKaye, 6 October 1926, MacKaye Family Papers; BM, *New Exploration*, 209–12; BM to Geoffrey Bolton, 9 March 1928, MacKaye Family Papers; BM to Clarence Stein, 15 November 1922, MacKaye Family Papers.

27. BM to Percy MacKaye, 6 October 1926; BM to Edward T. Hartman, 11 July 1927, both MacKaye Family Papers.

28. BM to Clarence Stein, 11 October 1921; BM to Clarence Stein, 26 February 1922; Clarence Stein to BM, 2 March 1922; BM to Clarence Stein, 10 February 1922; BM to Geoffrey Bolton, 9 March 1928, all MacKaye Family Papers.

29. Foresta, "Transformation of the Appalachian Trail," 79, 82; BM, "Wilderness Ways," 242.

30. BM to "Mrs. Floyd," 25 August 1925; BM to Frank E. Lutz, 13 March 1929, both MacKaye Family Papers; BM, "Flankline vs. Skyline"; Sutter, *Driven Wild*, 23, 29–30.

31. BM, "Wilderness Ways," 239; BM, "Appalachian Trail: A Guide," 330; BM, "Appalachian Trail: A Project"; Arthur C. Comey to BM, 24 September 1925, MacKaye Family Papers; BM to Lewis Mumford, 3 December 1926, MacKaye Family Papers.

32. BM to Clarence Stein, 14 February 1922, 6 October 1922; Paul M. Fink to BM, 14 February 1924, all MacKaye Family Papers.

33. Armitage, *Nature Study Movement*, 158–59; BM, "Outdoor Culture" speech, Northeast Trail Conference, Boston, 21 January 1927 (reprinted in *Landscape Architecture*, April 1927); Paul M. Fink to BM, 22 March 1924, MacKaye Family Papers; Anderson, *Benton MacKaye*, 161; Carr, *Wilderness by Design*, 11, 14; Meinig, *Interpretation of Ordinary Landscapes*, 38; Appalachian Trail Conference, Report of the Fifth Annual Meeting. Ethan Carr notes that manipulated images, including photomosaics and marked-up photographs, became standard practice for park and road engineers in the 1920s. See Carr, *Wilderness by Design*, 170.

34. BM to Harry Dow, 22 December 1928; BM to Clarence Stein, 14 February 1922, both MacKaye Family Papers; BM, "Appalachian Trail: A Project"; Luccarelli, "Benton MacKaye's Appalachian Trail," 213; Sutter, "Retreat from Profit," 565. Douglas Reichert Powell frames the AT, in MacKaye's hands, as a rhetorical practice. He pushed for a collective imagining in which a "simple" forest trail might illuminate a "complex relationship among places." See Powell, *Critical Regionalism*, 61.

35. Torrey, Place, and Dickinson, *New York Walk Book*, xiv; BM to Theodora Kimball Hubbard, 23 February 1927, MacKaye Family Papers; Raymond Torrey to BM, 17 June 1929, MacKaye Family Papers; BM to Raymond Torrey, 22 June 1929, 27 June 1929, MacKaye Family Papers.

36. Easterling, *Organizational Space*, 21; Ness, "Path Taken," 43; BM, "Appalachian Trail: A Guide," 339; BM to Clarence Stein, 31 August 1922, 15 September 1922, MacKaye Family Papers.

37. BM, "Appalachian Trail: A Guide," 332; BM, "Outdoor Culture"; BM, "Wilderness Ways," 249; Minteer, *Landscape of Reform*, 102; Carr, *Wilderness by Design*, 75; BM to H. G. Merriam, 7 April 1932, MacKaye Family Papers.

38. BM to Clarence Stein, 9 July 1932; BM to Hazel MacKaye, 27 July 1932, 4 September 1932, all MacKaye Family Papers.

39. BM to Hazel MacKaye, 27 July 1932, MacKaye Family Papers; BM, "The AT: A Guide to the Study of Nature" proposal, November 1932, MacKaye Family Papers.

40. BM, "The AT: A Guide to the Study of Nature" proposal; Dann, *Across the Great Border Fault*, 33.

41. BM to Hazel MacKaye, 4 September 1932, MacKaye Family Papers.

42. Paul M. Fink letter to the Seventh ATC, 22 June 1935, MacKaye Family Papers.

43. Anderson, *Benton MacKaye*, 306; Parsons, "Collaborative Genius," 466; Minteer, "Wilderness and the Wise Province," 199; Sutter, "Retreat from Profit," 570.

44. BM, "Outdoor Culture."

45. BM to Hazel MacKaye, 1 July 1934, MacKaye Family Papers.

46. BM to Lewis Mumford, 4 April 1927, MacKaye Family Papers; Sutter, *Driven Wild*, 142.

47. BM, "Regional Planning," 294–95.

Chapter 3 • The Stern Grip of Circumstance: Hiking the Smoky Mountains, 1920–1940

1. William Hart Jr., "George Masa: The Best Mountaineer," in *May We All Remember Well*, ed. Brunk, 249.

2. Powell, *Critical Regionalism*, 46–47.

3. Smoky Mountains Hiking Club (hereafter SMHC) 1927 handbook, HB box 1, SMHC Collection; Potomac Appalachian Trail Club, "Hiking, Camping, and Mountaineering Equipment" (1960), Appalachian Trail vertical file, Paul M. Fink Papers.

4. D. A. Brown, "Modern Romance of Mountaineering," 4, 24; SMHC 1935 Handbook, SMHC Handbook Digital Collection.

5. *Ruralite* [Sylva], NC, 29 June 1926; undated *Asheville Citizen* article (ca. 1930), George Masa biographical file, North Carolina Collection; 5 February 1931 meeting minutes, box 1, folder 1, Carolina Mountain Club Archives (hereafter CMCA); undated *Asheville Times* article, ca. 1930, "Find Auto Trip Not Practical in NC Area of Smoky NP," Masa scrapbook, George Masa Collection; Hart, "George Masa," in *May We All Remember Well*, ed. Brunk, 249, 255.

6. February 1961 article, *Asheville Times*, Wilderness Hikers Scrapbooks, Samuel Robinson Papers; George Masa to Fred Seely, 15 March 1932, North Carolina Collection. When Masa died, the CMC scrambled to obtain the thousands of negatives he had created over the years. In June 1935, the CMC passed a motion to buy all available Masa negatives for $30 or to spend $25 to buy prints from the negatives. The goal was to present the materials to the national park museum that was then being planned for the Smokies. See July 1933, August 1933, and June 1935 meeting minutes, box 1, folder 1, CMCA.

7. Tuan, *Space and Place*, 123–24; Eleanor Stephens, notes on photograph MA042-8, North Carolina Collection; undated *Asheville Citizen* article (ca. 1930), George Masa biographical file, North Carolina Collection; "Birds, Scenery, and Folks," 16; Hyde, *American Vision*, 82.

8. C. C. Campbell, *Memories of Old Smoky*, 18; C. C. Campbell, *Birth of a National Park*, 23, 25; Pierce, *Great Smokies*, 53.

9. C. C. Campbell, *Memories of Old Smoky*, 12, 72, 73.

10. C. C. Campbell, *Birth of a National Park*, 14; R. Campbell, *In Darkest Alaska*, 75.

11. *Kingsport Times*, 13 November 1935, 11 October 1936, 29 October 1936; Fink, *Backpacking Was the Only Way*, 63, 273.

12. Fink, *Backpacking Was the Only Way*, 147, 149, 153, 248–49; Faris, *Roaming the Eastern Mountains*, 273.

13. Bagley, *Aerophotography and Aerosurveying*, 109; Eardley, *Aerial Photographs*, 25, 75.

14. 31 March 1930 executive board meeting minutes, BDMIN box 1, SMHC Collection; 30 January 1934 board of directors meeting minutes, SMHC Collection; 11 January 1932 board of directors meeting minutes, BDMIN box 1, SMHC Collection; *Asheville Times*, 10 March 1930. The club's racial exclusion policy has been noted before. For a

group that seemed so unconventional in many respects, this was perhaps the most conventional of its outlooks. They were part of a national collection of wilderness advocates who "articulated the profound racism that characterized the era." See O'Brien, *Landscapes of Exclusion*, 30.

15. "The Carolina Mountain Club," folder 7, box 15, CMCA; CMC Bulletin, January 1932, CMCA; "Mountaineering Information," 1935, CMCA; "Here is Hikers' Paradise," undated newspaper article, box 9, CMCA; George Revill and Charles Watkins, "Educated Access: Interpreting Forestry Commission Forest Park Guides," in Watkins, ed., *Rights of Way*, 122–23.

16. ATC Publication no. 8, "Guide to the Southern Appalachians," 1937, SMHC Collection; 1935 SMHC handbook, HB box 1, SMHC Collection; April 1933 meeting minutes, box 1, folder 1, CMCA; April 1935 meeting minutes, box 1, folder 1, CMCA. For examples of the CMC requesting photos, see 5 February 1931 and 1 June 1931 meeting minutes, box 1, folder 1, CMCA.

17. Littlefield and Steere ad, *This Week in Knoxville and East Tennessee*, June 1930, East Tennessee Automobile Club Scrapbook.

18. Bowman, *Land of High Horizons*, 63; July 1929 *Asheville Times* article, Masa scrapbook, George Masa Collection; caption to Chimney Tops photograph in folder 6, box 2, Great Smoky Mountains National Park Photography Collection; *Knoxville Journal*, 22 February 1931; Broome, *Out under the Sky*, 41.

19. George Masa log, George Masa Collection; Shumaker, "Untold Stories"; Hart, "George Masa," in *May We All Remember Well*, ed. Brunk, 256–57, 261.

20. Fink, *Backpacking Was the Only Way*, 99.

21. Faris, *Roaming the Eastern Mountains*, 262.

22. R. Roberts, "Great Women," 54; Thornborough, "Tramping in the Great Smokies," 1; Thornborough, *Great Smoky Mountains*, 126, 130.

23. Thornborough, *Great Smoky Mountains*, 122; D. A. Brown, "Modern Romance of Mountaineering," 2, 4.

24. Gelber, *Hobbies*, 12.

25. Broome, "Introducing the Smoky Mountains Hiking Club," 19; J. S. Holmes to George Masa, quoted in Hart, "George Masa," in *May We All Remember Well*, ed. Brunk, 249–75. A third organization, the Wilderness Hikers of Asheville, formed in early 1939 and attracted fewer members. See Wilderness Hikers of Asheville constitution and bylaws, Wilderness Hikers Scrapbook, Samuel Robinson Papers; *Asheville Times*, 7 August 1939.

26. Kimberly A. Jarvis, "Gender and Wilderness Conservation," in *American Wilderness*, ed. M. Lewis, 162–63; SMHC Program of Hikes for 1928, SMHC Handbook Digital Collection; Tebbutt, "Rambling and Manly Identity," 1125–53; *Kingsport Times*, 28 October 1931; Fink, *Backpacking Was the Only Way*, 181.

27. Meeting minutes from 2 April 1931, 6 August 1931, 10 December 1931, 19 February 1932, December 1932, December 1933, January 1935, February 1935, April 1935, box 1, folder 1, CMCA; 30 May 1931 hike report, box 8, CMCA.

28. *Knoxville Journal*, 23 September 1933; BM to Hazel MacKaye, 21 September 1933 and 28 April 1934, MacKaye Family Papers; BM to Jake MacKaye, 6 November 1934, MacKaye Family Papers.

29. 1929 SMHC handbook, HB box 1, SMHC Collection.

30. Daugstad, "Negotiating Landscape," 406. Actual insiders—mountain residents—rarely factored into hiking accounts, though when they did, it was often in a mildly confrontational manner. Paul Fink, on a trip up Big Bald Mountain in 1916, ran into a "tall, lank native" who questioned him to see whether he was a government tax agent; Fink, *Backpacking Was the Only Way*, 23. A CMC trip to Soco Bald in August 1931 was interrupted by the brief appearance of "two real 'moonshiners' both heavily armed"; 2 August 1931 hike report, box 8, CMCA. The SMHC reported that many of its blazes were destroyed by gunshots; A. G. Roth, "Report of the SMHC at the Appalachian Trail Conference," August 1937, SMHC Collection.

31. Csikszentmihalyi, *Flow*, 97; Csikszentmihalyi, *Beyond Boredom and Anxiety*, 13.

32. SMHC Program of Hikes for 1930, SMHC Handbook Digital Collection; C. C. Campbell, *Memories of Old Smoky*, 21.

33. Csikszentmihalyi, *Beyond Boredom and Anxiety*, 30; McCoy and Masa, *Guide*, 98; Bowman, *Land of High Horizons*, 91; 1931 SMHC handbook, HB box 1, SMHC Collection; 1932 SMHC handbook, HB box 1, SMHC Collection.

34. Csikszentmihalyi, *Beyond Boredom and Anxiety*, 36; Fink, *Backpacking Was the Only Way*, 55; SMHC Program of Hikes for 1930, SMHC Handbook Digital Collection.

35. SMHC Handbook Digital Collection.

36. Broome, *Out under the Sky*, 11; Kropp, "Wilderness Wives," 6.

37. Broome, *Out under the Sky*, 29; Fink, *Backpacking Was the Only Way*, 128; Stoddard, *Shanks' Mare*, 142; Edensor, "Walking in the British Countryside," 101; Csikszentmihalyi, *Beyond Boredom and Anxiety*, 44.

38. Cresswell, *Place*, 10; ATC Publication 8, "Guide to the Southern Appalachians," 1937, 95–96; Fink, *Backpacking Was the Only Way*, 97.

39. Unattributed article, ca. August 1924 ("To Young People Who Love Smoky Mountains Goes Credit for Great Interest in the National Forest"), East Tennessee Automobile Club Scrapbook; Fink, *Backpacking Was the Only Way*, 184; *Knoxville News*, 12 August 1924; *Asheville Times*, 22 July 1929, 23 June 1931.

40. Jeffers, *Call of the Mountains*, 247; *Hickory Daily Record*, 5 September 1935; Fink, *Backpacking Was the Only Way*, 56. The art historian Diana Strazdes finds a similar relationship between stunning vistas and clogged, obstructed foregrounds in the work of the Hudson River School in the mid-nineteenth century. See Strazdes, "Wilderness and Its Waters," 350–51.

41. SMHC Program of Hikes for 1927, SMHC Handbook Digital Collection; Fink, *Backpacking Was the Only Way*, 11, 48.

42. Huntley, *Making of Yosemite*, 169; Williams et al., "Beyond the Commodity Metaphor," 41.

43. Kawamura, Suzuki, and Morikawa, "Effect of Expertise," 769, 774.

44. C. C. Campbell, *Memories of Old Smoky*, 21–33.

45. Cohen, *Pathless Way*, 71; K. Mitchell, *Spirituality and the State*, 63–65; Stoddard, *Shanks' Mare*, 13, 80; *New York Times*, 20 April 1924.

46. SMHC Handbook Digital Collection.

47. Judson, *Jack and Jill*, 35; Kephart, *Our Southern Highlanders*, 43–44, 136; Harvey Broome to J. Ross Eakin, 15 December 1931, COR box 1, SMHC Collection; Arno Cammerer to E. G. Frizzell, 30 December 1932, COR box 1, SMHC Collection; Fink, *Backpacking Was the Only Way*, 104. The aspiration for fluid walking styles echoes Edensor's observa-

tion that British guidebooks for "serious walkers" have stressed that the "body must be schooled in correct walking technique and the walker must persistently check the terrain for the requisite performance." See Edensor, "Walking in the British Countryside," 96–97.

48. SMHC 1933 Handbook, SMHC Handbook Digital Collection.

49. Fink, *Backpacking Was the Only Way*, 114.

50. C. C. Campbell, *Memories of Old Smoky*, 21–33; Stoddard, *Shanks' Mare*, 122.

51. SMHC Program of Hikes for 1929, SMHC Handbook Digital Collection; Stoddard, *Shanks' Mare*, 137.

52. Bowman, *Land of High Horizons*, 89–90; Judson, *Jack and Jill*, 10; 26 July 1932 board of directors meeting minutes, SMHC Collection; ATC Publication 8, "Guide to the Southern Appalachians," 1937, 100; Guy Frizzell to Robert White, 6 February 1934, COR box 1, SMHC Collection; J. R. Eakin to Arno Cammerer, 29 September 1936, COR box 1, SMHC Collection.

53. Broome, *Out under the Sky*, 12; Fink, *Backpacking Was the Only Way*, 29; *Asheville Times*, 9 July 1929; Paul M. Fink, "Why?," SMHC 1931 handbook, HB box 1, SMHC Collection; 3 April 1932 and 9 October 1932 hike reports, box 8, CMCA; SMHC 1927 handbook, HB box 1, SMHC Collection. See also 13 March 1932 hike report for a similar description of the top of Yates Knob.

54. Dacy, "Appalachian Trail," 4; CMC pamphlet, spring 1924, box 15, folder 24, CMCA; Fink, *Backpacking Was the Only Way*, 67; "Here is Hikers' Paradise," undated newspaper article, box 9, CMCA; SMHC Program of Hikes for 1928, SMHC Handbook Digital Collection.

55. Broome, *Out under the Sky*, 36; Fink, *Backpacking Was the Only Way*, 201.

56. CMC Trip Report, May 1932, George Masa Collection.

Chapter 4 • A Priceless Asset: Marion Pearsall in Isolation

1. Kephart, *Our Southern Highlanders*, 29.

2. Fraley, "Missionaries to the Wilderness," 35; Pearsall, "Some Behavioral Factors," 205.

3. Pearsall, "Healthways," 13.

4. Pearsall, *Little Smoky Ridge*, x, 3.

5. Pearsall, "Some Aspects of Cultural Change," 5; Strauss, "Marion Pearsall," 437–38; *Tuscaloosa News*, 12 August 1956.

6. Marion Pearsall to Wiley Faw, 11 April 1959, box 9, folder 9-204, Marion Pearsall Collection (hereafter MPC); Pearsall, *Little Smoky Ridge*, 27, 81, 106, 167; Elmore, "Field Notes," 46. The Kentucky Agricultural Experiment Station paid for Pearsall's 1959 research, and she sought outside funding for the following year. She received a $750 grant from the Wenner-Gren Foundation for Anthropological Research for the follow-up work. In the grant application, Pearsall explained the significance of her work by noting that she hoped to analyze "little understood factors of physical, social, and cultural accessibility operating against the adequate provision and use of health services in remote areas." Pearsall was careful to stress that "one rural, relatively isolated low-income county" was useful to researchers so far as it offered general findings that could inform action research. Grant application, folder 11-256, MPC; *Kentucky Kernel* [Lexington], 5 February 1960.

7. Pearsall, *Little Smoky Ridge*, vii, 2, 92, 97, 133; Gold, "Roles in Sociological Field Observations," 222.

8. Pearsall, "Healthways," 7.

9. Mary Breckinridge to Marion Pearsall, 20 April 1959, box 9, folder 9-204, MPC; West, "History, Organization," 218–35; Breckinridge, *Wide Neighborhoods*, 163; folder 26-616, MPC.

10. Redfield, *Little Community*, 7, 91.

11. Thomas Ford, afterword to J. S. Brown, *Beech Creek*, 286; "Research on Mental Health in Appalachia," Southern Regional Education Board report, folder 9-190, MPC.

12. Marion Pearsall to Edith Shaw, 5 April 1960, folder 9-204, MPC; June 1959 field notes, folder 9-209, MPC.

13. 1959 field notes, MPC; speech notes, 18 June 1959, folder 10-210, MPC.

14. Quarles, "Comparison of Some Aspects," 9; Kimball and Pearsall, "Event Analysis," 63.

15. Breckinridge, *Wide Neighborhoods*, 158.

16. 1959 field notes, MPC.

17. *Frontier Nursing Service Quarterly Bulletin* 48 (Summer 1972): 37.

18. Caudill, *Darkness at Dawn*, 15; Ford, *Southern Appalachian Region*, 18; Miller, *Approaches to University Extension Work*, 10.

19. Greenwood and Levin, *Introduction to Action Research*, 18; Alan Banks, Dwight Billings, and Karen Tice, "Appalachian Studies, Resistance, and Postmodernism," in *Fighting Back in Appalachia*, ed. Fisher, 294; Beers, "Highland Society in Transition," 27; Giffin, "From Cinder Hollow to Cincinnati," 16; Quarles, "Comparison of Some Aspects," 2.

20. Gold, "Roles in Sociological Field Observations," 218; Kimball and Pearsall, "Event Analysis," 59–60.

21. Kimball and Pearsall, "Event Analysis," 61, 63; Pearsall, "Some Behavioral Factors," 207.

22. Whisnant, *All That Is Native and Fine*, 264; Pearsall, "Some Behavioral Factors," 204; Appalachian Regional Commission, *Appalachian Kentucky*, 3, 38.

23. Harry Caudill, foreword to Weller, *Yesterday's People*, xiii; Gary C. Cox, "Topophilia and Technological Change in Appalachia," in *Sense of Place in Appalachia*, ed. Whitson, 253; Drake, "Recession Is Far from Over," 37.

24. Pearsall, "Healthways," 10; Pearsall, "Some Behavioral Factors," 203, 206; 1959 field notes, MPC.

25. Miller, *Approaches to University Extension Work*, 79; "Research on Mental Health in Appalachia," Southern Regional Education Board report, folder 9-190, Box 9, MPC.

26. Pearsall, "Healthways," 8, 9, 12–13.

27. Pearsall, "Healthways in a Southern Appalachian County" prospectus, folder 11-253, box 11, MPC. As late as 1974, Pearsall still considered this book an ongoing project. See *Frontier Nursing Service Quarterly Bulletin* 50 (Summer 1974): 17.

28. Pearsall, "Model for the Analysis," 212.

29. Ibid., 212–13, 215.

30. Pearsall, "Some Behavioral Factors," 200.

31. Pearsall, "Model for the Analysis," 215.

32. Quarles, "Comparison of Some Aspects," 48; 1964 Census of Agriculture, vol. 1, pt. 30: Kentucky, 81, 156, 287; Motter, "Time Marches On," 12–15.

33. Pearsall, "Some Behavioral Factors," 201; 1959 field notes, MPC; Quarles, "Comparison of Some Aspects," 2, 32.

34. Quarles, "Comparison of Some Aspects," 62, 136; 1959 field notes, MPC; Appalachian Regional Commission, *Appalachian Kentucky*, 3, 35–38; Ford, afterword to J. S. Brown, *Beech Creek*, 285.

35. Pearsall, "Healthways," 9; Pearsall, "Some Aspects of Culture Change," 150.

36. Rupert B. Vance, "The Region: A New Survey," in *Southern Appalachian Region*, ed. Ford, 13; Pearsall, "Some Behavioral Factors," 201; Weller, *Yesterday's People*, 26; June 1959 field notes, MPC; L. Roberts, *Up Cutshin and Down Greasy*, vii.

37. 1959 field notes, MPC; March 1960 field notes, MPC; Pearsall, "Some Behavioral Factors," 201.

38. Manderscheid, "Movement Problem," 606; Böhm et al., *Against Automobility*, 3.

39. Brooks, "Strip Mining in East Kentucky," 125; Quarles, "Comparison of Some Aspects," 33; June 1959 field notes, MPC; 1960 field notes, MPC.

40. 1959 field notes, MPC.

41. Ibid.

42. Ibid.; 1960 field notes, MPC.

43. 1960 field notes, MPC.

44. Ibid.

45. Ibid.

46. Quarles, "Comparison of Some Aspects," 67; April 1960 field notes, MPC; 1959 field notes, MPC.

47. Pearsall, "Healthways," 10.

48. 1959 field notes, MPC; 1960 field notes, MPC.

49. *Thousandsticks* [Hyden, KY], 7 April 1960; 1959 field notes, MPC.

50. 1959 field notes, MPC; Pearsall, "Some Behavioral Factors," 205.

51. 1959 field notes, MPC.

52. Pearsall, "Model for the Analysis," 215.

53. Stephenson, *Shiloh*, 226.

54. Kernodle, "Last of the Rugged Individualists," 46–51.

55. Marion Pearsall to Mary Breckenridge, 15 April 1959, box 9, folder 9-204, MPC; Marion Pearsall to John Bischoff, 24 February 1960, folder 9-204, MPC; Marion Pearsall to Mary Breckinridge, 5 April 1960, folder 9-204, MPC; *Kentucky Kernel* [Lexington], 5 March 1960.

56. Eller, *Uneven Ground*, 54; Bradshaw, *Appalachian Regional Commission*, 30, 40; Walls and Stephenson, *Appalachia in the Sixties*, xi; June 1959 field notes, MPC.

57. Ford, *Southern Appalachian Region*, 11, 298.

58. Weller, *Yesterday's People*, viii, 13, 24.

59. Pearsall, "Participant Observation," 38.

60. Miller, *Approaches to University Extension Work*, 123; Merton, "Selected Problems," 305; Trice, "'Outsider's' Role," 29, 31; Pearsall, "Participant Observation," 39.

61. Pearsall, "Some Behavioral Factors," 208.

62. Trice, "'Outsider's' Role," 29; Weller, *Yesterday's People*, 3; Eller, *Uneven Ground*, 63.

Chapter 5 • William Gedney and the Look of Coal Country

1. For examples of these photographers, see Buchanan and Hawthorne, *Picture Man*; and Lentz, *W. R. Trivett*.

2. Wilma Cornett Johnson, unknown photographer, undated photograph, Facebook, August 4, 2011, https://www.facebook.com/wilma.johnson.58.

3. Harry Ernst and Charles Drake, "The Lost Appalachians," *Nation*, 30 May 1959, in *Appalachia in the Sixties*, ed. Walls and Stephenson, 5.

4. Between 1950 and 1970, almost twelve thousand Leslie County residents, fifty-three thousand Harlan County residents, and thirty-seven thousand Perry County residents moved away. "Miscellaneous Writings," William Gedney Photographs and Writings (hereafter WGPW); Eller, *Uneven Ground*, 20, 28; Bradshaw, *Appalachian Regional Commission*, 20; Weller, *Yesterday's People*, 21; Appalachian Regional Commission, *Appalachian Kentucky*, 3, 8; Ford, "Social Change in Eastern Kentucky," 11; Dan Wakefield, "In Hazard," *Commentary* (September 1963), in *Appalachia in the Sixties*, ed. Walls and Stephenson, 11; Drake, "Recession Is Far from Over," 36–37; Bowman and Haynes, *Resources and People*, 15; Bigart, "Kentucky Miners," 1.

5. Batteau, *Invention of Appalachia*, 155, 161; Wakefield, "In Hazard," in *Appalachia in the Sixties*, ed. Walls and Stephenson, 11; *New York Times*, 13 January 1952, 54; *New York Times*, 7 August 1953, 16; Bigart, "Depressed Kentucky Coal Area," 1.

6. Bigart, "Kentucky Miners," 79; Bigart, "Depressed Kentucky Coal Area," 1.

7. Bowman and Haynes, *Resources and People*, 32, 35.

8. Bradshaw, *Appalachian Regional Commission*, 30; Eller, *Uneven Ground*, 81; Associated Press, "Johnson Visits Shacks in Appalachian Tour," 25 April 1964; Associated Press, "Wonderful Spirit Seen by Johnson in Poverty Areas," 25 April 1964; *New York Times*, 10 May 1964; Ford, "Social Change in Eastern Kentucky," 5; Bigart, "Kentucky Miners," 1.

9. *New York Times*, 19 December 1963; *New York Times*, 26 December 1963.

10. Steven H. Scheuer, syndicated review of "Depressed Area, U.S.A.," February 22, 1964; Kiffmeyer, *Reformers to Radicals*, 100.

11. Associated Press, "Santa Comes to Appalachia by Plane and Money Order," 24 December 1964; Batteau, *Invention of Appalachia*, 7.

12. Murphy, "Valley of Poverty," 54–65; Batteau, *Invention of Appalachia*, 161; Weller, *Yesterday's People*, 87.

13. Sartor and Dyer, *What Was True*, 20; "Miscellaneous Writings," WGPW.

14. "Miscellaneous Writings," WGPW.

15. Ibid.

16. Ibid.

17. *Hazard Herald*, 4 March 1965.

18. "Miscellaneous Writings," WGPW; Dyer, *Otherwise Known*, 61; Scully, "Seeing Pictures," 50; Projects Notebook, WGPW.

19. Information about Gedney's composition of the shot is from undeveloped negatives on contact sheet no. 1016, WGPW.

20. "Miscellaneous Writings," July 4, 1971, WGPW; Weller, *Yesterday's People*, 98.

21. Thrift, "Intensities of Feeling," 75; *New York Times*, 20 October 1963.

22. Weller, *Yesterday's People*, 88.

23. "Miscellaneous Writings," WGPW.

24. 1970 Photography Memo Book, WGPW.

25. 1969 Notebook, WGPW.

26. Hoffer, *Working and Thinking*; India Daily Diary, WGPW.

27. 1970 Photography Memo Book, WGPW.

28. "Miscellaneous Writings," July 23, 1971, WGPW; Trillin, "Logical Thing, Costwise," 111; Transcriptions and Notes, 1972, WGPW.

29. Travel Notes 1975, WGPW.

30. Ibid.

31. Thompson, *Truth and Photography*, 91.

32. Thomas Ford, "Passing of Provincialism," in *Southern Appalachian Region*, ed. Ford, 9; Ford, "Social Change in Eastern Kentucky," 2.

33. Travel Notes 1975, WGPW.

34. "Miscellaneous Writing," WGPW.

35. Travel Notes 1975, WGPW.

36. "Miscellaneous Writings," WGPW.

37. Transcriptions and Notes, 27 December 1973, WGPW; Cameron, "When Strangers Bring Cameras," 413; Thompson, *Truth and Photography*, 96.

38. Sartor and Dyer, *What Was True*, 21; Hostetler, *Street Seen*, 22.

39. Sartor and Dyer, *What Was True*, 21–22; Museum of Modern Art press release, 8 February 1968, MOMA Press Release Archive; Tucker, "Lyons, Szarkowski," 25; "Six Win Guggenheim Awards," 12.

40. Gedney Guggenheim application, WGPW; Cameron, "When Strangers Bring Cameras," 412.

41. David Chaney, "The Power of Metaphors in Tourism Theory," in *Tourism*, ed. Coleman and Crang, 204; Gedney Guggenheim application, WGPW.

42. *New York Times*, 15 July 1962, 74.

43. Westerbeck and Meyerowitz, *Bystander*, 403. For Szarkowski's views on these photographers, see MOMA press releases, 5 December 1974, 7 July 1966, 21 February 1963, 28 February 1967, 28 February 1967, and 26 September 1974, respectively, MOMA Press Release Archive.

44. Hoffer, *Working and Thinking*, 169; Memo Book, February 1968–August 1969, WGPW.

45. Rathbone, *Walker Evans*, 264, 266; Westerbeck and Meyerowitz, *Bystander*, 280; Sartor and Dyer, *What Was True*, 30; Evans, *Many Are Called*; Evans, *Message from the Interior*.

46. "Thirty-One Photographs," WGPW; Rathbone, *Walker Evans*, 160; Thompson, *Truth and Photography*, 89; 1967–1969 Notebook, WGPW.

47. Szarkowski, "Different Kind of Art," 64.

48. MOMA press release, 28 February 1967, MOMA Press Release Archive.

49. MOMA wall label, 17 December 1968–10 March 1969, MOMA Press Release Archive.

50. Coleman, "Latent Image," 21; MOMA Press Release, 17 December 1968, MOMA Press Release Archive; Hostetler, *Street Seen*, 165.

51. Westerbeck and Meyerowitz, *Bystander*, 381; Snyder and Allen, "Photography, Vision, and Representation," 165; Tucker, "Lyons, Szarkowski," 27.

52. Mora, *Last Photographic Heroes*, 128; "Interview with John Szarkowski," 2; John Simon Guggenheim Memorial Foundation, *Reports of the President*, 147.

53. Sartor and Dyer, *What Was True*, 26; 1969 Notebook, WGPW; Scully, "Seeing Pictures," 50, 52.

54. 1967–1969 Notebook, WGPW.

55. Szarkowski, "Different Kind of Art," 68; Tucker, "Lyons, Szarkowski," 28; 1969 Notebook, WGPW.

56. 1973 Roll Book, WGPW.

57. Photography Technical Book, WGPW; 1973 Roll Book, WGPW.

58. 1973 Roll Book, WGPW; Travel Notes, 1975, WGPW; Kennedy, "Stubbornly Practicing His Principles"; Heise, "Hugh Edwards."

59. NLRB v. UMW 8280, 422 F.2d 115, No. 18742 (6th Cir. 1969); Vivian Cornett to William Gedney, 12 August 1964, box 1, WGPW; Vivian Cornett to William Gedney, 2 January 1965 and 18 May 1965, WGPW.

60. Transcriptions and Notes, WGPW; Vivian Cornett to William Gedney, 18 May 1965, WGPW.

61. Boyd Couch to William Gedney, 6 January 1965, WGPW; *Hazard Herald*, 4 March 1965.

62. "Interview with John Szarkowski," 3; Ledford, "Thanks for Not Shooting Me," 395; Transcriptions and Notes, 27 December 1973, WGPW.

63. "Miscellaneous Writings," 5 June 1972, WGPW; Vivian Cornett to William Gedney, 10 June 1972, WGPW.

64. Vivian Cornett to William Gedney, 25 October 1972; "Miscellaneous Writings," box 1, WGPW.

65. Transcriptions and Notes, 27 December 1973, WGPW; "Miscellaneous Writings," August 1971, WGPW.

Conclusion

1. Vaughn, "Purpose of This Magazine," 2–3; Mitchell and Schnyder, "Public Relations for Appalachia."

2. Vaughn, "Purpose of This Magazine," 3; Berry, Obermiller, and Scott, *Studying Appalachian Studies*, 5.

3. Scott, *Removing Mountains*, 222.

4. Penny Messinger, roundtable response in "Epilogue: Reflections on the Concept of Place," in *Women of the Mountain South*, ed. Rice and Tedesco, 478; Scott, *Removing Mountains*, 14; Alan Banks, Dwight Billings, and Karen Tice, "Appalachian Studies, Resistance, and Postmodernism," in *Fighting Back in Appalachia*, ed. Fisher, 294.

5. Schumann and Fletcher, *Appalachia Revisited*, 14. On the diversity of Appalachian Studies approaches, see Engelhardt, *Beyond Hill and Hollow*; Rice and Tedesco, *Women of the Mountain South*; and Bell, *Fighting King Coal*.

6. Steve Fuller, "The Constitutively Social Character of Expertise," in *Philosophy of Expertise*, ed. Selinger and Crease, 347.

7. Obermiller and Maloney, "Uses and Misuses of Appalachian Culture."

8. Crang, "Picturing Practices," 367; Matless, Watkins, and Merchant, "Nature Trails," 98.

Primary Sources

Manuscript Collections

Carolina Mountain Club Archives. D. H. Ramsey Library Special Collections. University of North Carolina–Asheville. Asheville, NC.

George Masa Collection. Hunter Library Special Collections. Western Carolina University. Cullowhee, NC.

Great Smoky Mountains National Park Photography Collection. Great Smoky Mountains National Park Library and Archive. Gatlinburg, TN.

J. Horace McFarland Papers. Pennsylvania State Archives. Harrisburg, PA.

MacKaye Family Papers. Rauner Special Collections Library. Dartmouth College. Hanover, NH.

Marion Pearsall Collection. Special Collections Research Center. University of Kentucky. Lexington, KY.

Museum of Modern Art Press Release Archive. www.moma.org/research-and-learning/research-resources/press_archives/.

North Carolina Collection. Pack Memorial Library. Asheville, NC.

Paul M. Fink Papers. W. L. Eury Appalachian Collection. Appalachian State University. Boone, NC.

Samuel Robinson Papers. D. H. Ramsey Library Special Collections. University of North Carolina at Asheville. Asheville, NC.

Smoky Mountains Hiking Club Collection. McClung Historical Collection. Knox County Public Library. Knoxville, TN.

William Gedney Photographs and Writings. David M. Rubenstein Rare Book & Manuscript Library. Duke University. Durham, NC.

Published Works

"Abuse of the Optical Lantern." *Photo-Era* 26 (January 1911): 26.

Appalachian Regional Commission. *Appalachian Kentucky: Past and Promise*. Lexington, KY: Secretary for Development, 1975.

"At the Men's Convention in Harrisburg, Pa." *Religious Telescope* 74 (18 November 1908): 8.

Bagley, James W. *Aerophotography and Aerosurveying*. New York: McGraw-Hill, 1941.

Beardsley, A. H. "The Automobile and the Beginner." *Photo-Era* 43 (November 1919): 264–65.

Beers, Howard W. "Highland Society in Transition." *Mountain Life and Work* 22 (Spring 1946): 1–3, 26.

Bigart, Homer. "Depressed Kentucky Coal Area Fears Killings in Labor Unrest." *New York Times*, 6 December 1962.

——. "Kentucky Miners: A Grim Winter." *New York Times*, 20 October 1963.

"Birds, Scenery, and Folks." *Outing* 71 (October 1917): 16.

Bowman, Elizabeth Skaggs. *Land of High Horizons*. Kingsport, TN: Southern, 1938.

Bowman, Mary Jean, and W. Warren Haynes. *Resources and People in East Kentucky: Problems and Potentials of a Lagging Economy*. Baltimore: Johns Hopkins Press, 1963.

Breckinridge, Mary. *Wide Neighborhoods: A Story of the Frontier Nursing Service*. New York: Harper & Brothers, 1952.

Brooks, David B. "Strip Mining in East Kentucky." *Mountain Life and Work* 43 (Spring 1967): 4–9.

Broome, Harvey. "Introducing the Smoky Mountains Hiking Club." *Appalachian Trailway News* 9 (May 1948).

——. *Out under the Sky of the Great Smokies: A Personal Journey*. Privately printed, 1975. Reprint, Knoxville: University of Tennessee Press, 2001.

Brown, James S. *Beech Creek: A Study of a Kentucky Mountain Neighborhood*. Berea, KY: Berea College Press, 1988.

Buttrick, P. L. "Tree-Planting Plans for Michigan's Highways." *American City* 23 (July 1920): 11–12.

Campbell, Carlos C. *Birth of a National Park in the Great Smoky Mountains*. Knoxville: University of Tennessee Press, 1960.

——. *Memories of Old Smoky: Early Experiences in the Great Smoky Mountains*. Privately published, 1967. Reprint, Knoxville: University of Tennessee Press, 2005.

Caudill, Harry M. *A Darkness at Dawn: Appalachian Kentucky and the Future*. Lexington: University Press of Kentucky, 1976.

Claudy, C. H. "The Lantern at Home." *Photo-Era* 26 (February 1911): 64.

Coleman, A. D. "Latent Image." *Village Voice*, 16 January 1969, 21.

Dacy, George H. "The Appalachian Trail." *Mentor* 16 (1928): 1–14.

Drake, Chad. "The Recession Is Far from Over in the Southern Mountains." *Mountain Life and Work* 34 (Winter 1958): 36–37.

"Dr. J. Horace McFarland." *Recreation* 42 (March 1949): 559.

Eardley, A. J. *Aerial Photographs: Their Use and Interpretation*. New York: Harper & Brothers, 1941.

Elmore, Peggy. "Field Notes." *Frontier Nursing Service Quarterly Bulletin* 46 (Autumn 1970): 44–50.

Evans, Walker. *Many Are Called*. Boston: Houghton Mifflin, 1966.

——. *Message from the Interior*. New York: Eakins, 1966.

Faris, John T. *Roaming the Eastern Mountains*. New York: Farrar & Rinehart, 1932.

Field, Parker. "The Appalachian Mountain Club." *Photo-Era* 52 (April 1924): 190.

Fink, Paul M. *Backpacking Was the Only Way: A Chronicle of Camping Experiences in the Southern Appalachian Mountains*. Johnson City: East Tennessee State University, 1975.

Ford, Thomas. "Social Change in Eastern Kentucky since 1960." University of Kentucky Center for Developmental Change, paper no. 10, 1973.

——, ed. *The Southern Appalachian Region: A Survey*. Lexington: University of Kentucky Press, 1962.

Giffin, Roscoe. "From Cinder Hollow to Cincinnati." *Mountain Life and Work* 32 (Winter 1956): 11–20.

Gold, Raymond L. "Roles in Sociological Field Observations." *Social Forces* 36 (March 1958): 217–23.

"Hand Cameras." *Outing* 35 (October 1899): 100.

"Harvard's Course in Printing." *Inland Printer* 47 (September 1911): 913.

"He Helped to Save Niagara Falls." *Hampton's Magazine* 24 (June 1909): 864–65.

Hoffer, Eric. *Working and Thinking on the Waterfront: A Journal, June 1958–May 1959*. New York: Harper & Row, 1969.

"An Interview with John Szarkowski." *MOMA* 7 (Summer 1978): 2.

Jeffers, Le Roy. *The Call of the Mountains: Rambles among the Mountains and Canyons of the United States and Canada*. New York: Dodd, Mead, 1922.

"J. Horace McFarland." *Inland Printer* 48 (December 1911): 436.

John Simon Guggenheim Memorial Foundation. *Reports of the President and the Treasurer*. 1966.

Kephart, Horace. *Our Southern Highlanders*. 2nd ed. New York: Macmillan, 1922.

Kernodle, Rigdon Wayne. "Last of the Rugged Individualists." *Harper's*, January 1960, 46–51.

Kimball, Solon T., and Marion Pearsall. "Event Analysis as an Approach to Community Study." *Social Forces* 34 (October 1955): 58–63.

"Lantern Slides." *Outing* 35 (October 1899): 100.

Lee, W. S. "Snapshots." *Photo-Era* 43 (November 1919): 254.

MacKaye, Benton. "Appalachian Power: Servant or Master." *Survey Graphic* 51 (1 March 1924): 618–19.

——. "The Appalachian Trail: A Guide to the Study of Nature." *Scientific Monthly* 34 (April 1932): 330–42.

——. "An Appalachian Trail: A Project in Regional Planning." *Journal of the American Institute of Architects* 9 (October 1921): 325–30.

——. "A Boys' Project for a Balanced Civilization." *Christian Science Monitor*, 13 January 1927.

——. "End or Peak of Civilization?" *Survey* 68 (1932): 441–44.

——. "Flankline vs. Skyline." *Appalachia* 20 (1934): 104–8.

——. *The New Exploration: A Philosophy of Regional Planning*. New York: Harcourt, Brace, 1928.

——. "Progress toward the Appalachian Trail." *Appalachia* 15 (December 1922): 244–45.

——. "Regional Planning." *Sociological Review* 20 (October 1928): 293–99.

——. "Wilderness Ways." *Landscape Architecture* 19 (July 1929): 237–49.

McCoy, George W., and George Masa. *Guide to the Great Smoky Mountains National Park*. 1st ed. Asheville, NC: Inland, 1933.

McFarland, J. Horace. "Are National Parks Worth While?" *Proceedings of the National Parks Session of the American Civic Association*. Washington, DC: Government Printing Office, 1911.

——. "The Awakening of a City." *World's Work* 3 (April 1902): 1930–32.
——. "Bringing the Folks to the Forest." *Review of Reviews*, July 1929, 132–33.
——. *Eagles Mere and the Sullivan Highlands*. Harrisburg, PA: J. Horace McFarland, 1944.
——. *Floral Designs*. Philadelphia: A. Blanc & J. H. McFarland, 1888.
——. *Getting Acquainted with the Trees*. New York: Outlook, 1904.
——. "The Highest Development of the Summer Camp." *Suburban Life* 7 (July 1908): 18–20.
——. "Horticultural Photography." In *Standard Cyclopedia of Horticulture*, 2600. New York: Macmillan, 1919.
——. "Hunting in the May Woods." *Outlook* 83 (26 May 1906): 190–95.
——. "The Ignorance of 'Good' Citizens." *Outlook* 82 (3 February 1906): 270–73.
——. *Photographing Flowers and Trees and the Use of Natural Forms in Decorative Photography*. New York: Tennant & Ward, 1911.
——. "The Romance of Color Photography." *Suburban Life* 9 (December 1909): 298, 336–37.
——. "Shall Our Country Roads Be Beautiful or Ugly?" *Country Calendar* 1 (August 1905): 369.
——. "Shall We Have Ugly Conservation?" *Outlook* 91 (13 March 1909): 594–98.
——. "Shall We Make a Coal-Pile of Niagara?" *Ladies Home Journal* 22 (September 1905): 19.
——. "Statement in Regard to the Lecture on Color Photography." *Journal of the Engineers Society of Pennsylvania* 1 (June 1909): 259.
——. "A Summer Lake Carnival." *Suburban Life* 9 (July 1909): 9.
——. "Twenty Years of Scenery-Saving in America." *Landscape Architecture* 20 (July 1930): 301–7.
——. "What about the Country Highway?" *National Municipal Review* 10 (April 1921): 221–25.
——. "Why Billboard Advertising as at Present Conducted Is Doomed." *Chautauquan* 51 (June 1908): 19–46.
Merton, Robert K. "Selected Problems of Field Work in the Planned Community." *American Sociological Review* 12 (June 1947): 304–12.
Miller, Robert. *Approaches to University Extension Work with the Rural Disadvantaged*. Morgantown, WV: Office of Research and Development, Appalachian Center, 1972.
Motoring with a Kodak. Eastman Kodak, 1910.
Motter, Margaret. "Time Marches On in the Hinterlands." *Mountain Life and Work* 24 (Fall 1948): 12–15.
Murphy, Michael. "The Valley of Poverty." *Life*, 31 January 1964, 54–65.
"Outdoor Backgrounds." *Photo-Era* 43 (September 1919): 145.
Pearsall, Marion. "Healthways in a Mountain Community." *Mountain Life and Work* 36 (1960): 7–13.
——. *Little Smoky Ridge: The Natural History of a Southern Appalachian Neighborhood*. Birmingham: University of Alabama Press, 1959.
——. "A Model for the Analysis of Cross-Cultural Action Programs." *Human Organization* 19 (Winter 1960): 212–15.

——. "Participant Observation as Role and Method in Behavioral Research." *Nursing Research* 14 (Winter 1965): 37–42.

——. "Some Aspects of Culture Change in a Mountain Neighborhood of East Tennessee." PhD diss., University of California, Berkeley, 1950.

——. "Some Behavioral Factors in the Control of Tuberculosis in a Rural County." *American Review of Respiratory Disease* 85 (1962): 200–210.

"Print Criticism." *Photo-Era* 26 (January 1911): 37.

Quarles, Mary Ann Stillman. "A Comparison of Some Aspects of Family Life between Two Areas of Leslie County, Kentucky." MA thesis, University of Kentucky, 1952.

Redfield, Robert. *The Little Community: Viewpoints for the Study of a Human Whole*. Chicago: University of Chicago Press, 1955.

Roberts, Leonard. *Up Cutshin and Down Greasy: Folkways of a Kentucky Mountain Family*. Lexington: University of Kentucky Press, 1959.

Scully, Julia. "Seeing Pictures." *Modern Photography*, April 1969, 50–52.

"Six Win Guggenheim Awards." *Popular Photography* 59 (May 1966): 12.

Snyder, Joel, and Neil Walsh Allen. "Photography, Vision, and Representation." *Critical Inquiry* 2 (Autumn 1975): 143–69.

Stephenson, John. *Shiloh: A Mountain Community*. Lexington: University Press of Kentucky, 1968.

Stoddard, Charles Coleman. *Shanks' Mare*. New York: George H. Doran, 1924.

Szarkowski, John. "A Different Kind of Art." *New York Times Magazine*, 13 April 1975, 64.

Thornborough, Laura. *The Great Smoky Mountains*. New York: Thomas Crowell, 1937.

——. "Tramping in the Great Smokies." *American Forests and Forest Life* 33 (August 1927): 463–66, 512.

Torrey, Raymond, Frank Place Jr., and Robert Dickinson. *New York Walk Book*. New York: American Geographical Society, 1923.

Trice, Harrison. "The 'Outsider's' Role in Field Study." *Sociology and Social Research* 41 (September 1956): 27–32.

Trillin, Calvin. "The Logical Thing, Costwise." *New Yorker*, 27 December 1969, 33.

Vaughn, Marshall Everett. "Purpose of This Magazine." *Mountain Life and Work* 1 (1 April 1925): 2–3.

Watrous, Richard B. "Civic Art and Country Life." *Annals of the American Academy of Political and Social Science* 40 (March 1912): 191–99.

Weller, Jack E. *Yesterday's People: Life in Contemporary Appalachia*. Lexington: University of Kentucky Press, 1965.

Secondary Sources

Anderson, Larry. *Benton MacKaye: Conservationist, Planner, and Creator of the Appalachian Trail*. Baltimore: Johns Hopkins University Press, 2002.

——. "Benton MacKaye's 1904 White Mountains Hike: Exploring a Landscape of Logging, 'Camp Ethics,' and Patriotism." In *Nature and Culture in the Northern Forest: Region, Heritage, and Environment in the Rural Northeast*, ed. Pavel Cenkl, 153–70. Iowa City: University of Iowa Press, 2010.

Armitage, Kevin C. *The Nature Study Movement: The Forgotten Popularizer of America's Conservation Ethic*. Lawrence: University Press of Kansas, 2009.

Batteau, Allen W. *The Invention of Appalachia*. Tucson: University of Arizona Press, 1990.

Bell, Claudia, and John Lyall, *The Accelerated Sublime: Landscape, Tourism, and Identity*. Westport, CT: Praeger, 2002.

Bell, Shannon Elizabeth. *Fighting King Coal: The Challenges to Micromobilization in Central Appalachia*. Lexington: University Press of Kentucky, 2016.

Berleant, Arnold. *The Aesthetics of Environment*. Philadelphia: Temple University Press, 1992.

Berry, Chad, Phillip J. Obermiller, and Shaunna L. Scott, eds. *Studying Appalachian Studies: Making the Path by Walking*. Urbana: University of Illinois Press, 2015.

Billings, Dwight B., Gurney Norman, and Katherine Ledford, eds. *Confronting Appalachian Stereotypes: Back Talk from an American Region*. Lexington: University Press of Kentucky, 1999.

Blair, Bob, ed. *William Henry Jackson's "The Pioneer Photographer."* Santa Fe: Museum of New Mexico Press, 2005.

Böhm, Steffen, Campbell Jones, Chris Land, and Matthew Paterson, eds. *Against Automobility*. Malden, MA: Blackwell, 2006.

Bradshaw, Michael. *The Appalachian Regional Commission: Twenty-Five Years of Government Policy*. Lexington: University Press of Kentucky, 1992.

Brown, Douglas A. "The Modern Romance of Mountaineering: Photography, Aesthetics and Embodiment." *International Journal of the History of Sport* 24 (January 2007): 1–34.

Brunk, Robert S., ed. *May We All Remember Well*. Vol. 1. Asheville, NC: Brunk Auction Services, 1997.

Buchanan, Paul, and Ann Hawthorne. *The Picture Man*. Chapel Hill: University of North Carolina Press, 1993.

Cameron, Ardis. "When Strangers Bring Cameras: The Poetics and Politics of Othered Places." *American Quarterly* 54 (September 2002): 411–35.

Campbell, Robert. *In Darkest Alaska: Travel and Empire along the Inside Passage*. Philadelphia: University of Pennsylvania Press, 2008.

Carr, Ethan. *Wilderness by Design: Landscape Architecture and the National Park Service*. Lincoln: University of Nebraska Press, 1998.

Chamberlin, Silas. *On the Trail: A History of American Hiking*. New Haven, CT: Yale University Press, 2016.

Chambliss, Julian C. "Perfecting Space: J. Horace McFarland and the American Civic Association." *Pennsylvania History* 77 (Autumn 2010): 486–97.

Cohen, Michael P. *The Pathless Way: John Muir and American Wilderness*. Madison: University of Wisconsin Press, 1984.

Coleman, Simon, and Mike Crang, eds. *Tourism: Between Place and Performance*. New York: Berghahn, 2002.

Cosgrove, Denis, and Veronica della Dora. *High Places: Cultural Geographies of Mountains, Ice, and Science*. London: I. B. Tauris, 2009.

Crang, Mike. "Picturing Practices: Research through the Tourist Gaze." *Progress in Human Geography* 21 (1997): 359–74.

Cresswell, Tim. *Place: A Short Introduction*. Malden, MA: Blackwell, 2004.

Cronon, William. "The Trouble with Wilderness; or, Getting Back to the Wrong Nature." In *Uncommon Ground: Rethinking the Human Place in Nature*, 69–90. New York: W. W. Norton, 1995.

Csikszentmihalyi, Mihaly. *Beyond Boredom and Anxiety: The Experience of Play in Work and Games*. San Francisco: Jossey-Bass, 1975.

——. *Flow: The Psychology of Optimal Experience*. New York: Harper & Row, 1990.

Cutlip, Scott. *Public Relations History: From the 17th to the 20th Century, the Antecedents*. Hillsdale, NJ: Lawrence Erlbaum Associates, 1995.

Dann, Kevin. *Across the Great Border Fault: The Naturalist Myth in America*. New Brunswick, NJ: Rutgers University Press, 2000.

Daugstad, Karoline. "Negotiating Landscape in Rural Tourism." *Annals of Tourism Research* 35 (2008): 402–26.

Dear, Peter. "Mysteries of State, Mysteries of Nature: Authority, Knowledge, and Expertise in the Seventeenth Century." In *States of Knowledge: The Co-production of Science and Social Order*, ed. Sheila Jasanoff, 206–24. London: Routledge, 2004.

Dunlap, Thomas R. *Faith in Nature: Environmentalism as Religious Quest*. Seattle: University of Washington Press, 2004.

Dunn, Durwood. *Cades Cove: The Life and Death of a Southern Appalachian Community, 1818–1937*. Knoxville: University of Tennessee Press, 1988.

Dyer, Geoff. *Otherwise Known as the Human Condition: Selected Essays and Reviews*. Minneapolis, MN: Graywolf, 2011.

Easterling, Keller. *Organizational Space: Landscapes, Highways, and Houses in America*. Cambridge, MA: MIT Press, 1999.

Edensor, Tim. "Walking in the British Countryside: Reflexivity, Embodied Practices and Ways to Escape." *Body and Society* 6 (November 2000): 81–106.

Eller, Ronald D. *Uneven Ground: Appalachia since 1945*. Lexington: University Press of Kentucky, 2008.

Engelhardt, Elizabeth S. D., ed. *Beyond Hill and Hollow: Original Readings in Appalachian Women's Studies*. Athens: Ohio University Press, 2005.

Enright, Kelly. *The Maximum of Wilderness: The Jungle in the American Imagination*. Charlottesville: University of Virginia Press, 2012.

Entrikin, Nicolas. *The Betweenness of Place: Toward a Geography of Modernity*. Baltimore: Johns Hopkins University Press, 1991.

Featherstone, Mike, Nigel Thrift, and John Urry, eds. *Automobilities*. London: SAGE, 2005.

Feintuch, Burt. "The End of an Era of Innocence." *Journal of Appalachian Studies* 16 (Spring/Fall 2010): 105.

Feld, Steven, and Keith Basso, eds. *Senses of Place*. Santa Fe, NM: School of American Research Press, 1996.

Fisher, Stephen L., ed. *Fighting Back in Appalachia: Traditions of Resistance and Change*. Philadelphia: Temple University Press, 1993.

Foresta, Ronald. "Transformation of the Appalachian Trail." *Geographical Review* 77 (January 1987): 76–85.

Fraley, Jill. "Missionaries to the Wilderness: A History of Land, Identity, and Moral Geography in Appalachia." *Journal of Appalachian Studies* 17 (Spring/Fall 2011): 28–41.

Gelber, Steven. *Hobbies: Leisure and the Culture of Work in America*. New York: Columbia University Press, 1999.

Greenwood, Davydd, and Morten Levin, *Introduction to Action Research: Social Research for Social Change*. Thousand Oaks, CA: SAGE, 2007.

Haney, David H. *When Modern Was Green: Life and Work of Landscape Architect Leberecht Migge*. London: Routledge, 2010.

Hays, Samuel P. *Conservation and the Gospel of Efficiency: The Progressive Conservation Movement, 1890–1920*. Cambridge, MA: Harvard University Press, 1959.

Heise, Kenan. "Hugh Edwards, Friend to Art of Photography." *Chicago Tribune*, 27 August 1986.

Hostetler, Lisa. *Street Seen: The Psychological Gesture in American Photography, 1940–1959*. Munich: Prestel, 2009.

Hou, Shen. *The City Natural:* Garden and Forest *Magazine and the Rise of American Environmentalism*. Pittsburgh, PA: University of Pittsburgh Press, 2013.

House, Silas, and Jason Howard. *Something's Rising: Appalachians Fighting Mountaintop Removal*. Lexington: University Press of Kentucky, 2009.

Hsiung, David C. *Two Worlds in the Tennessee Mountains: Exploring the Origins of Appalachian Stereotypes*. Lexington: University Press of Kentucky, 1997.

Huntley, Jen A. *The Making of Yosemite: James Mason Hutchings and the Origin of America's Most Popular National Park*. Lawrence: University Press of Kansas, 2011.

Hyde, Anne Farrar. *An American Vision: Far Western Landscape and National Culture, 1820–1920*. New York: New York University Press, 1990.

Jenkins, Reese V. *Images and Enterprise: Technology and the American Photographic Industry 1839 to 1925*. Baltimore: Johns Hopkins University Press, 1975.

Johnson, Jeffrey Guy. "Feud, Society, Family: Feud Narratives in the United States, 1865–1910." PhD diss., Harvard University, 2000.

Jordan-Bychkov, Terry G. *The Upland South: The Making of an American Folk Region and Landscape*. Santa Fe, NM: Center for American Places, 2003.

Judd, Richard. *Common Lands, Common People: The Origins of Conservation in Northern New England*. Cambridge, MA: Harvard University Press, 1997.

Judson, John C. *Jack and Jill Go Up a Hill*. Gatlinburg, TN: Little Pigeon, 1940.

Kawamura, Satoru, Sae Suzuki, and Kazunori Morikawa. "The Effect of Expertise in Hiking on Recognition Memory for Mountain Scenes." *Memory* 15 (October 2007): 768–75.

Kennedy, Randy. "Stubbornly Practicing His Principles of Photography." *New York Times*, 26 April 2009.

Kiffmeyer, Thomas. *Reformers to Radicals: The Appalachian Volunteers and the War on Poverty*. Lexington: University Press of Kentucky, 2008.

Knight, Jan. "The Environmentalism of Edward Bok: The *Ladies' Home Journal*, the General Federation of Women's Clubs, and the Environment, 1901–09." *Journalism History* 29 (Winter 2004): 154–65.

Kolodny, Annette. *The Lay of the Land: Metaphor as Experience and History in American Life and Letters*. Chapel Hill: University of North Carolina Press, 1975.

Kropp, Phoebe. "Wilderness Wives and Dishwashing Husbands: Comfort and the Domestic Arts of Camping in America, 1880–1910." *Journal of Social History* 43 (2009): 5–30.

Lamme, Margot Opdycke. "The 'Public Sentiment Building Society.'" *Journalism History* 29 (Fall 2003): 123–32.

Lamme, Margot Opdycke, and Karen Miller Russell. "Removing the Spin: Toward a New Theory of Public Relations History." *Journalism and Communication Monographs* 11 (Winter 2010): 281–362.

Ledford, Katherine. "Thanks for Not Shooting Me." *Appalachian Journal* 31 (Spring 2004): 390–98.

Lentz, Ralph E. *W. R. Trivett, Appalachian Pictureman: Photographs of a Bygone Time*. Jefferson, NC: McFarland, 2001.

Lewis, Michael, ed. *American Wilderness: A New History*. New York: Oxford University Press, 2007.

Lewis, Ronald L. *Transforming the Appalachian Countryside: Railroads, Deforestation, and Social Control in West Virginia, 1880–1920*. Chapel Hill: University of North Carolina, 1998.

Li, Tania Murray. *The Will to Improve: Governmentality, Development, and the Practice of Politics*. Durham, NC: Duke University Press, 2007.

Luccarelli, Mark. "Benton MacKaye's Appalachian Trail: Imagining and Engineering a Landscape." In *Technologies of Landscape: From Reaping to Recycling*, ed. David E. Nye, 207–17. Amherst: University of Massachusetts Press, 1999.

Manderscheid, Katharina. "The Movement Problem, the Car, and Future Mobility Regimes: Automobility as Dispositif and Mode of Regulation." *Mobilities* 9 (4): 604–26.

Maoz, Darya. "The Mutual Gaze." *Annals of Tourism Research* 33 (January 2006): 221–39.

Marx, Leo. *The Machine in the Garden: Technology and the Pastoral Ideal in America*. New York: Oxford University Press, 1967.

Matless, David, Charles Watkins, and Paul Merchant. "Nature Trails: The Production of Instructive Landscapes in Britain, 1960–72." *Rural History* 21 (April 2010): 97–131.

Meinig, D.W., ed. *The Interpretation of Ordinary Landscapes*. New York: Oxford University Press, 1979.

Minteer, Ben A. *The Landscape of Reform: Civic Pragmatism and Environmental Thought in America*. Cambridge, MA: MIT Press, 2006.

——. "Wilderness and the Wise Province: Benton MacKaye's Pragmatic Vision." *Philosophy and Geography* 4 (2001): 185–202.

Mitchell, Catherine C., and C. Joan Schnyder. "Public Relations for Appalachia: Berea's *Mountain Life and Work*." *Journalism Quarterly* 66 (December 1989): 974–1049.

Mitchell, Don. "The Lure of the Local: Landscape Studies at the End of a Troubled Century." *Progress in Human Geography* 25 (2001): 269–82.

Mitchell, Kerry. *Spirituality and the State: Managing Nature and Experience in America's National Parks*. New York: New York University Press, 2016.

Mitchell, Robert D., ed. *Appalachian Frontiers: Settlement, Society, and Development in the Preindustrial Era*. Lexington: University Press of Kentucky, 1991.

Mom, Gijs. "Civilized Adventure as a Remedy for Nervous Times: Early Automobilism and *Fin-de-siècle* Culture." *History of Technology* 23 (2001): 157–90.

Mora, Gilles. *The Last Photographic Heroes: American Photographers of the Sixties and Seventies*. New York: Harry N. Abrams, 2007.

Morin, Karen. *Frontiers of Femininity: A New Historical Geography of the Nineteenth-Century American West*. Syracuse, NY: Syracuse University Press, 2008.

Morrison, Ernest. *J. Horace McFarland: A Thorn for Beauty*. Harrisburg: Pennsylvania Historical and Museum Commission, 1995.

Nash, Roderick. *Wilderness and the American Mind*. 3rd ed. New Haven, CT: Yale University Press, 1982.

Ness, Erik. "The Path Taken." *Preservation* 55 (July/August 2003): 40–43.

Nye, David E. *American Technological Sublime*. Cambridge, MA: MIT Press, 1994.
Obermiller, Phillip J., and Michael E. Maloney. "The Uses and Misuses of Appalachian Culture." *Journal of Appalachian Studies* 22 (Spring 2016): 103–12.
O'Brien, William E. *Landscapes of Exclusion: State Parks and Jim Crow in the American South*. Amherst: University of Massachusetts Press, 2016.
Orvell, Miles. *The Real Thing: Imitation and Authenticity in American Culture, 1880–1940*. Chapel Hill: University of North Carolina Press, 1989.
Parker, Gavin. "The Country Code and the Ordering of Countryside Citizenship." *Journal of Rural Studies* 22 (January 2006): 1–16.
Parsons, Kermit. "Collaborative Genius: The Regional Planning Association of America." *Journal of the American Planning Association* 60 (Autumn 1994): 462–82.
Peterson, Jon A. "The Birth of Organized City Planning in the United States, 1909–1910." *Journal of the American Planning Association* 75 (Spring 2009): 123–33.
Pierce, Daniel S. *The Great Smokies: From Natural Habitat to National Park*. Knoxville: University of Tennessee Press, 2000.
Powell, Douglas Reichert. *Critical Regionalism: Connecting Politics and Culture in the American Landscape*. Chapel Hill: University of North Carolina Press, 2007.
Rathbone, Belinda. *Walker Evans: A Biography*. Boston: Houghton Mifflin, 1995.
Reid, Herbert G. "Appalachia and the 'Sacrament of Co-existence': Beyond Post-Colonial Trauma and Regional Identity Traps." *Journal of Appalachian Studies* 11 (Spring/Fall 2005): 164–81.
Rice, Connie Park, and Marie Tedesco. *Women of the Mountain South: Identity, Work, and Activism*. Athens: Ohio University Press, 2015.
Rimby, Susan. *Mira Lloyd Dock and the Progressive Era Conservation Movement*. State College: Pennsylvania State University Press, 2012.
Roberts, Rachel. "Great Women of the Great Smoky Mountains National Park: Defining Nature and the National Park Service." MA thesis, Appalachian State University, 2012.
Rose, Gillian. *Feminism and Geography: The Limits to Geographical Knowledge*. Minneapolis: University of Minnesota Press, 1993.
Russell, Dave. *Looking North: Northern England and the National Imagination*. Manchester: Manchester University Press, 2004.
Saito, Yuriko. "The Aesthetics of Unscenic Nature." *Journal of Aesthetics and Art Criticism* 56 (Spring 1998): 101–11.
Sartor, Margaret, and Geoff Dyer, eds. *What Was True: The Photographs and Notebooks of William Gedney*. New York: W. W. Norton, 2000.
Satterwhite, Emily. "Imagining Home, Nation, World: Appalachia on the Mall." *Journal of American Folklore* 121 (Winter 2008): 10–34.
Schrepfer, Susan. *Nature's Altars: Mountains, Gender, and American Environmentalism*. Lawrence: University Press of Kansas, 2005.
Schumann, William, and Rebecca Adkins Fletcher, eds. *Appalachia Revisited: New Perspectives on Place, Tradition, and Progress*. Lexington: University Press of Kentucky, 2016.
Scott, Rebecca R. *Removing Mountains: Extracting Nature and Identity in the Appalachian Coalfields*. Minneapolis: University of Minnesota Press, 2010.
Selinger, Evan, and Robert P. Crease, eds. *The Philosophy of Expertise*. New York: Columbia University Press, 2006.

Shapiro, Henry D. *Appalachia on Our Mind: The Southern Mountains and Mountaineers in the American Consciousness, 1870–1920*. Chapel Hill: University of North Carolina Press, 1978.

Shields, Rob. *Places on the Margin: Alternative Geographies of Modernity*. London: Routledge, 1991.

Shumaker, Susan. "Untold Stories from America's National Parks: George Masa." www.pbs.org/nationalparks/media/pdfs/tnp-abi-untold-stories-pt-06-masa.pdf.

Simo, Melanie L. *Forest and Garden: Traces of Wildness in a Modernizing Land, 1897–1949*. Charlottesville: University of Virginia Press, 2003.

Sørensen, Ninna Nyberg, and Finn Stepputat. "Narrations of Authority and Mobility." *Identities* 8 (2001): 313–42.

Stehr, Nico, and Reiner Grundmann. *Experts: The Knowledge and Power of Expertise*. London: Routledge, 2011.

Stilgoe, John R. *Landscape and Images*. Charlottesville: University of Virginia Press, 2005.

Stokowski, Patricia A. "Languages of Place and Discourses of Power: Constructing New Senses of Place." *Journal of Leisure Research* 34 (2002): 368–82.

Stradling, David. *Making Mountains: New York City and the Catskills*. Seattle: University of Washington Press, 2007.

Strauss, Robert. "Marion Pearsall." *American Anthropologist* 88 (June 1986): 437–38.

Strazdes, Diana. " 'Wilderness and Its Waters': A Professional Identity for the Hudson River School." *Early American Studies* 7 (2009): 333–62.

Stroud, Ellen. *Nature Next Door: Cities and Trees in the American Northeast*. Seattle: University of Washington Press, 2012.

Sutter, Paul S. *Driven Wild: How the Fight against Automobiles Launched the Modern Wilderness Movement*. Seattle: University of Washington Press, 2002.

———. " 'A Retreat from Profit': Colonization, the Appalachian Trail, and the Social Roots of Benton MacKaye's Wilderness Advocacy." *Environmental History* 4 (October 1999): 553–77.

Taylor, John. *A Dream of England: Landscape, Photography, and the Tourist's Imagination*. Manchester: Manchester University Press, 1994.

Tebbutt, Melanie. "Rambling and Manly Identity in Derbyshire's Dark Peak, 1880s–1920s." *Historical Journal* 49 (December 2006): 1125–53.

Thompson, Jerry L. *Truth and Photography: Notes on Looking and Photographing*. Chicago: Ivan R. Dee, 2003.

Thrift, Nigel. "Intensities of Feeling: Towards a Spatial Politics of Affect." *Geografiska Annaler* 86 B (March 2004): 57–78.

Tuan, Yi-Fu. *Space and Place: The Perspective of Experience*. Minneapolis: University of Minnesota Press, 1977.

Tucker, Anne Willes. "Lyons, Szarkowski, and the Perception of Photography." *American Art* 21 (Fall 2007): 25–29.

Turner, James Morton. "From Woodcraft to 'Leave No Trace': Wilderness, Consumerism, and Environmentalism in Twentieth-Century America." *Environmental History* 7 (July 2002): 462–84.

Urry, John. *The Tourist Gaze*. 2nd ed. London: SAGE, 2002.

Van Slyck, Abigail Ayres. *A Manufactured Wilderness: Summer Camps and the Shaping of American Youth, 1890–1960*. Minneapolis: University of Minnesota Press, 2006.

Walls, David S., and John B. Stephenson, eds. *Appalachia in the Sixties: Decade of Reawakening*. Lexington: University Press of Kentucky, 1972.

Watkins, Charles, ed. *Rights of Way: Policy, Culture, and Management*. London: Pinter, 1996.

West, Edith. "History, Organization and the Changing Culture of Care: A Historical Analysis of the Frontier Nursing Service." *Journal of International Women's Studies* 14 (January 2013): 218–35.

Westerbeck, Colin, and Joel Meyerowitz. *Bystander: A History of Street Photography*. New York: Little, Brown, 1994.

Whisnant, David E. *All That Is Native and Fine: The Politics of Culture in an American Region*. Chapel Hill: University of North Carolina Press, 1983.

——. *Modernizing the Mountaineer: People, Power, and Planning in Appalachia*. Rev. ed. Knoxville: University of Tennessee Press, 1994.

White, Richard. "From Wilderness to Hybrid Landscapes: The Cultural Turn in Environmental History." *Historian* 66 (September 2004): 557–64.

Whitson, S. Mont, ed. *Sense of Place in Appalachia*. Morehead, KY: Morehead State University, 1988.

Wiebe, Robert H. *The Search for Order, 1877–1920*. New York: Hill & Wang, 1967.

Williams, Daniel R., Michael E. Patterson, Joseph W. Roggenbuck, and Alan E. Watson. "Beyond the Commodity Metaphor: Examining Emotional and Symbolic Attachment to Place." *Leisure Sciences* 14 (1992): 29–46.

Williams, John Alexander. "Appalachian History: Regional History in the Post-Modern Zone." *Appalachian Journal* 28 (Winter 2001): 168–87.

Wilson, William H. "Harrisburg's Successful City Beautiful Movement, 1900–1915." *Pennsylvania History* 47 (July 1980): 213–33.

——. "J. Horace McFarland and the City Beautiful Movement." *Journal of Urban History* 7 (May 1981): 315–34.

Young, Terence. "Social Reform through Parks: The American Civic Association's Program for a Better America." *Journal of Historical Geography* 22 (October 1996): 460–72.

INDEX

Page numbers in italics refer to figures.